Application of Photoactive Nanomaterials in Degradation of Pollutants

Application of Photoactive Nanomaterials in Degradation of Pollutants

Special Issue Editor

Roberto Comparelli

MDPI • Basel • Beijing • Wuhan • Barcelona • Belgrade

Special Issue Editor
Roberto Comparelli
National Research Council-Institute for Physical Chemical Processes (CNR-IPCF)
Italy

Editorial Office
MDPI
St. Alban-Anlage 66
4052 Basel, Switzerland

This is a reprint of articles from the Special Issue published online in the open access journal *Materials* (ISSN 1996-1944) from 2018 to 2019 (available at: https://www.mdpi.com/journal/materials/special_issues/photoactive_nanomaterials)

For citation purposes, cite each article independently as indicated on the article page online and as indicated below:

LastName, A.A.; LastName, B.B.; LastName, C.C. Article Title. *Journal Name* **Year**, *Article Number*, Page Range.

ISBN 978-3-03921-381-8 (Pbk)
ISBN 978-3-03921-382-5 (PDF)

Contents

About the Special Issue Editor

Roberto Comparelli (degree in Chemistry, 2001) received his Ph.D. in the Chemistry of Innovative Materials at University of Bari (Italy) in 2004. He is currently Staff Researcher at Italian National Research Council—Institute for Physical and Chemical Processes, Bari Division, Italy (CNR-IPCF). His expertise covers nanocrystal synthesis by wet chemistry (photoactive or magnetic oxides, II–VI semiconductors, metals) and their characterization (TEM, SEM, AFM, FT-IR, UV–Vis–NIR, PL) and surface engineering. He has a strong background in the application of photoactive nanocrystals in the degradation of organic/inorganic pollutants in water and gas matrices, and in the preparation and characterization of NC-based self-cleaning coatings. He is also interested in NC incorporation in polymer matrices, and applications in optoelectronic, self-assembly, biological, and environmental fields. He has been involved in several EU and Italian projects. He is the scientific director of research contracts with multinational companies aimed at the synthesis and characterization of innovative nanomaterials. He has co-authored over 100 papers (including 85 JCR publications, 8 book chapters, 3 international patents, and 2 Italian patents) in addition to having also co-authored over 200 contributions to National and International Congresses and Symposia and presenting invited talks. He is a member of the Editorial Board of *Crystals* (MDPI, ISSN 2073-4352), *Journal of Chemistry* (Hindawi Publishing Group, ISSN online: 2090-9071; doi:10.1155/2962), *Journal of Nanostructure in Chemistry* (Springer, ISSN online: 2193-8865), and member of the Advisory Board of *Sci* (MDPI, ISSN 2413-4155). He has also served as Guest Editor for several Special Issues published in peer-reviewed scientific journals.http://www.cnr.it/people/roberto.comparelli.

 *materials*

Editorial

Special Issue: Application of Photoactive Nanomaterials in Degradation of Pollutants

Roberto Comparelli

CNR-IPCF, Istituto Per i Processi Chimici e Fisici, S.S. Bari, c/o Dip. Chimica Via Orabona 4, 70126 Bari, Italy; roberto.comparelli@cnr.it

Received: 31 July 2019; Accepted: 1 August 2019; Published: 2 August 2019

Abstract: Photoactive nanomaterials are receiving increasing attention due to their potential application to light-driven degradation of water and gas-phase pollutants. However, to exploit the strong potential of photoactive materials and access their properties require a fine tuning of their size/shape dependent chemical-physical properties and on the ability to integrate them in photo-reactors or to deposit them on large surfaces. Therefore, the synthetic approach, as well as post-synthesis manipulation could strongly affect the final photocatalytic properties of nanomaterials. The potential application of photoactive nanomaterials in the environmental field includes the abatement of organic pollutant in water, water disinfection, and abatement of gas-phase pollutants in outdoor and indoor applications.

Keywords: photocatalysis; nanomaterials; advanced oxidation processes; water treatments; recalcitrant pollutants; gas-phase pollutants; NO_x; VOCs; building materials; disinfection

1. Introduction

In recent years, one of the most important concerns of the scientific community and society has been health and environmental protection via the smart and sustainable use of natural resources. In this context, water resources are gaining increasing attention due to the occurrence of emerging pollutants including dyes, pharmaceutical and personal care products, endocrine disruptors, and pathogens [1,2]. Moreover, the increasing amount of atmospheric pollutants has been regarded among the main causes of respiratory diseases such as emphysema and bronchitis, which arise from the contact of NOx and lungs [3]. Unfortunately, conventional pollution remediation methods show limited performances. For instance, in the field of water treatment adsorption or coagulation, such methods aim to concentrate pollutants by transferring them to other phases for example, sedimentation, filtration, chemical, and membrane technologies involve high operating costs and can generate toxic secondary pollutants in the ecosystem [4] and chlorination, although widely used in disinfection processes, can generate by-products associated with cancer or other pathologies [5].

It turns out that the interest of the scientific community has been focusing on alternative methods such as the "advanced oxidation processes (AOPs)". AOPs are convenient and innovative alternatives to conventional wastewater and air treatment processes aiming to accomplish the complete mineralization of organic pollutants (i.e., their conversion into safe by-products such as O_2, H_2O, N_2, and mineral acids) [6,7].

Among AOPs, semiconductor-based photocatalysis has recently emerged as a promising air/water treatment [8]. Photocatalysis takes place upon the activation of a semiconductor with electromagnetic radiation from sun or artificial light. When exposed to electromagnetic radiation, a semiconductor absorbs photons with sufficient energy to inject electrons from the valence band (VB) to its conduction band (CB), generating electron hole pairs (e^-/h^+). The h^+ have an electrochemical potential sufficiently positive to generate $^\bullet OH \cdot$ radicals from water or to directly oxidize many organic pollutants adsorbed onto the semiconductor surface, while the e^- react with oxygen molecules to form superoxide anions, $^\bullet O_2^-$, that quickly react with H^+ to finally produce $^\bullet OH$ radicals after a series of concatenated reactions in water or can directly reduce target molecules adsorbed on the surface [9].

Nanostructured catalysts have demonstrated improved performances with respect to its bulk counterpart, thanks to their extremely high surface-to-volume ratio that turns into a high density of catalytically active surface sites. In addition, due to the size-dependent band gap of nanosized semiconductors, it is possible to fine tune the redox potentials of photogenerated electron–hole pairs to selectively control photochemical reactions. Furthermore, charges photogenerated in nanocatalysts can easily reach the catalyst surface, thus decreasing the probability of bulk recombination [10]. Nonetheless, a large-scale application of nanosized photocatalysts for environmental purpose is still hampered by technological issues and by high costs related to its capability to obtain photoactive nanocatalysts with a high reaction yield and adequate morphological and structural control [11].

The aim of the present special issue is to report on recent progress towards the application of photoactive nanomaterials and nanomaterials-based coatings in pollutants degradation, paying particular attention to cases of study close to real application: Scalable synthetic approaches to nanocatalysts, preparation of nanocatalyst-based coatings, degradation of real pollutants and bacteria inactivation, and application in building materials.

2. This Special Issue

This special issue consists of one review and six research articles. The review from Petronella et al. reports a selection of synthetic approaches suitable for a large-scale production of mesoporous TiO_2-based photocatalysts. Attention has been focused on mesoporous TiO_2 due to its unique features, which include a high specific surface area, improved ultraviolet (UV) radiation absorption, high density of surface hydroxyl groups, and a significant ability for further surface functionalization. The overviewed synthetic strategies have been selected and classified according to the following criteria: (i) High reaction yield, (ii) reliable synthesis scale-up, and (iii) adequate control over morphological, structural, and textural features. The potential environmental applications of such nanostructures include water remediation and air purification which are also discussed [12].

The research article from Professor Mascolo and co-workers demonstrates the effectiveness of the novel multiphasic hydroxyapatite–TiO_2 material (HApTi) for the photocatalytic treatment of diclofenac. Diclofenac is one of the most detected pharmaceuticals in environmental water matrices and it is recalcitrant to conventional wastewater treatment plants. The authors investigated the toxicity of transformation products by using different assays: Daphnia magna acute toxicity test, Toxi-Chromo Test, and the *Lactuca sativa* and *Solanum lycopersicum* germination inhibition test. Overall, the toxicity of the samples obtained from the photocatalytic experiment with HApTi decreased at the end of the treatment, showing the potential applicability of the catalyst for the removal of diclofenac and the detoxification of water matrices [13].

Professor Yurdakal's group demonstrated the synthesis of Pt-loaded TiO_2 nanotube on Ti anode by anodic oxidation in ethylene glycol. Such an approach allowed the control of the length of the nanotube as a function of anodic oxidation time. The obtained materials were exploited in photoelectrocatalytic, electrocatalytic, and photocataytic degradation of Paraquat, one of the most widely used herbicides. The obtained results evidenced that the photoanodes show a significant synergy for photoelectrocatalytic activity [14].

A simple and low-cost method to preparing hybrid photocatalysts of copper (I) oxide/titania is proposed in the paper by Professor Kowalska and co-workers. They investigated the photocatalytic and antimicrobial properties of prepared nanocomposites in three reaction systems: Ultraviolet-visible (UV-Vis) induced methanol dehydrogenation and oxidation of acetic acid, and 2-propanol oxidation under visible light irradiation. Furthermore, bactericidal and fungicidal properties of Cu_2O/TiO_2 materials were analyzed under UV, visible, and solar irradiation, as well as for dark conditions [15].

Cu$_x$O thin films deposited using HiPIMS (high-power impulse magnetron sputtering) on polyester under different sputtering energies were successfully synthesized by Professor Rtimi and co-workers. The photocatalytic performance of the photocatalyst was evaluated for the degradation of a toxic textile dye (Reactive Green 12; RG12) under visible light LEDs irradiation. The recycling of the catalyst showed a high stability of the catalyst up to 21 RG12 discoloration cycles. ICP-MS showed stable ions' release after the 5th cycle for both ions. This allows potential industrial applications of the reported HiPIMS coatings in future [16].

Rehman et al. investigated the effects of TiO$_2$ nanoparticles on the sulfate attack resistance of ordinary Portland cement (OPC) and slag-blended mortars. The results show that the addition of nano-TiO$_2$ accelerated expansion, variation in mass, loss of surface microhardness, and widened cracks in OPC and slag-blended mortars. Nano-TiO$_2$ containing slag-blended mortars were more resistant to sulfate attack than nano-TiO$_2$ containing OPC mortars [17].

The research article from Liang et al. describes the synthesis and characterization of ZnO-ZnS core-shell nanorods by combining the hydrothermal method and vacuum sputtering. The results of comparative degradation efficiency toward methylene blue showed that the ZnO-ZnS nanorods with the shell thickness of approximately 17 nm had the highest photocatalytic performance. The highly stable catalytic efficiency and superior photocatalytic performance supports their potential for environmental applications [18].

Acknowledgments: I would like to acknowledge Louise Liu, Fannie Xu, and all the staff of the Materials Editorial Office for their great support during the preparation of this Special Issue. I would also like to thank all the authors for their great contributions, and the reviewers for the time they dedicated to reviewing the manuscripts.

Conflicts of Interest: The author declares no conflict of interest.

References

1. Ebele, A.J.; Abdallah, M.A.-E.; Harrad, S. Pharmaceuticals and personal care products (PPCPs) in the freshwater aquatic environment. *Emerg. Contam.* **2017**, *3*, 1–16. [CrossRef]

2. Li, J.; Liu, H.; Chen, J.P. Microplastics in freshwater systems: A review on occurrence, environmental effects, and methods for microplastics detection. *Water Res.* **2018**, *137*, 362–374. [CrossRef] [PubMed]

3. Balbuena, J.; Cruz-Yusta, M.; Sánchez, L. Nanomaterials to Combat NO(x) Pollution. *J. Nanosci. Nanotechnol.* **2015**, *15*, 6373–6385. [CrossRef] [PubMed]

4. Marques, J.A.; Costa, P.G.; Marangoni, L.F.B.; Pereira, C.M.; Abrantes, D.P.; Calderon, E.N.; Castro, C.B.; Bianchini, A. Environmental health in southwestern Atlantic coral reefs: Geochemical, water quality and ecological indicators. *Sci. Total Environ.* **2018**, *651*, 261–270. [CrossRef] [PubMed]

5. Fiorentino, A.; Ferro, G.; Alferez, M.C.; Polo-López, M.I.; Fernández-Ibañez, P.; Rizzo, L. Inactivation and regrowth of multidrug resistant bacteria in urban wastewater after disinfection by solar-driven and chlorination processes. *J. Photochem. Photobiol. B* **2015**, *148*, 43–50. [CrossRef] [PubMed]

6. Rizzo, L.; Malato, S.; Antakyali, D.; Beretsou, V.G.; Đolić, M.B.; Gernjak, W.; Heath, E.; Ivancev-Tumbas, I.; Karaolia, P.; Lado Ribeiro, A.R.; et al. Consolidated vs new advanced treatment methods for the removal of contaminants of emerging concern from urban wastewater. *Sci. Total Environ.* **2019**, *655*, 986–1008. [CrossRef]

7. Ângelo, J.; Andrade, L.; Madeira, L.M.; Mendes, A. An overview of photocatalysis phenomena applied to NOx abatement. *J. Environ. Manag.* **2013**, *129*, 522–539. [CrossRef]

8. Truppi, A.; Petronella, F.; Placido, T.; Striccoli, M.; Agostiano, A.; Curri, M.; Comparelli, R. Visible-Light-Active TiO$_2$-Based Hybrid Nanocatalysts for Environmental Applications. *Catalysts* **2017**, *7*, 100. [CrossRef]

9. Herrmann, J.-M. Photocatalysis fundamentals revisited to avoid several misconceptions. *Appl. Catal. B* **2010**, *99*, 461–468. [CrossRef]

10. Petronella, F.; Truppi, A.; Ingrosso, C.; Placido, T.; Striccoli, M.; Curri, M.L.; Agostiano, A.; Comparelli, R. Nanocomposite materials for photocatalytic degradation of pollutants. *Catal. Today* **2017**, *281 Pt 1*, 85–100. [CrossRef]

11. Petronella, F.; Curri, M.L.; Striccoli, M.; Fanizza, E.; Mateo-Mateo, C.; Alvarez-Puebla, R.A.; Sibillano, T.; Giannini, C.; Correa-Duarte, M.A.; Comparelli, R. Direct growth of shape controlled TiO$_2$ nanocrystals onto SWCNTs for highly active photocatalytic materials in the visible. *Appl. Catal. B* **2015**, *178*, 91–99. [CrossRef]

12. Petronella, F.; Truppi, A.; Dell'Edera, M.; Agostiano, A.; Curri, M.L.; Comparelli, R. Scalable Synthesis of Mesoporous TiO_2 for Environmental Photocatalytic Applications. *Materials* **2019**, *12*, 1853. [CrossRef] [PubMed]

13. Murgolo, S.; Moreira, I.S.; Piccirillo, C.; Castro, P.M.L.; Ventrella, G.; Cocozza, C.; Mascolo, G. Photocatalytic Degradation of Diclofenac by Hydroxyapatite–TiO_2 Composite Material: Identification of Transformation Products and Assessment of Toxicity. *Materials* **2018**, *11*, 1779. [CrossRef] [PubMed]

14. Özcan, L.; Mutlu, T.; Yurdakal, S. Photoelectrocatalytic Degradation of Paraquat by Pt Loaded TiO_2 Nanotubes on Ti Anodes. *Materials* **2018**, *11*, 1715. [CrossRef] [PubMed]

15. Janczarek, M.; Endo, M.; Zhang, D.; Wang, K.; Kowalska, E. Enhanced Photocatalytic and Antimicrobial Performance of Cuprous Oxide/Titania: The Effect of Titania Matrix. *Materials* **2018**, *11*, 2069. [CrossRef] [PubMed]

16. Zeghioud, H.; Assadi, A.A.; Khellaf, N.; Djelal, H.; Amrane, A.; Rtimi, S. Photocatalytic Performance of CuxO/TiO_2 Deposited by HiPIMS on Polyester under Visible Light LEDs: Oxidants, Ions Effect, and Reactive Oxygen Species Investigation. *Materials* **2019**, *12*, 412. [CrossRef] [PubMed]

17. Qudoos, A.; Kim, H.; Ryou, J.S. Influence of Titanium Dioxide Nanoparticles on the Sulfate Attack upon Ordinary Portland Cement and Slag-Blended Mortars. *Materials* **2018**, *11*, 356.

18. Liang, Y.-C.; Lo, Y.-R.; Wang, C.-C.; Xu, N.-C. Shell Layer Thickness-Dependent Photocatalytic Activity of Sputtering Synthesized Hexagonally Structured ZnO-ZnS Composite Nanorods. *Materials* **2018**, *11*, 87. [CrossRef] [PubMed]

 *materials* 

Review

Scalable Synthesis of Mesoporous TiO$_2$ for Environmental Photocatalytic Applications

Francesca Petronella [1], Alessandra Truppi [1], Massimo Dell'Edera [1,2], Angela Agostiano [1,2], M. Lucia Curri [1,2,*] and Roberto Comparelli [1,*]

[1] CNR-IPCF, Istituto Per i Processi Chimici e Fisici, U.O.S. Bari, c/o Dip. Chimica Via Orabona 4, 70126 Bari, Italy; f.petronella@ba.ipcf.cnr.it (F.P.); a.truppi@ba.ipcf.cnr.it (A.T.); m.delledera@ba.ipcf.cnr.it (M.D.'E.); angela.agostiano@uniba.it (A.A.)

[2] Università degli Studi di Bari "A. Moro", Dip. Chimica, Via Orabona 4, 70126 Bari, Italy

* Correspondence: lucia.curri@ba.ipcf.cnr.it (M.L.C.); roberto.comparelli@cnr.it (R.C.); Tel.: +39-080-5442027 (R.C.)

Received: 11 April 2019; Accepted: 5 June 2019; Published: 7 June 2019

Abstract: Increasing environmental concern, related to pollution and clean energy demand, have urged the development of new smart solutions profiting from nanotechnology, including the renowned nanomaterial-assisted photocatalytic degradation of pollutants. In this framework, increasing efforts are devoted to the development of TiO$_2$-based nanomaterials with improved photocatalytic activity. A plethora of synthesis routes to obtain high quality TiO$_2$-based nanomaterials is currently available. Nonetheless, large-scale production and the application of nanosized TiO$_2$ is still hampered by technological issues and the high cost related to the capability to obtain TiO$_2$ nanoparticles with high reaction yield and adequate morphological and structural control. The present review aims at providing a selection of synthetic approaches suitable for large-scale production of mesoporous TiO$_2$-based photocatalysts due to its unique features including high specific surface area, improved ultraviolet (UV) radiation absorption, high density of surface hydroxyl groups, and significant ability for further surface functionalization The overviewed synthetic strategies have been selected and classified according to the following criteria (i) high reaction yield, (ii) reliable synthesis scale-up and (iii) adequate control over morphological, structural and textural features. Potential environmental applications of such nanostructures including water remediation and air purification are also discussed.

Keywords: photocatalysis; titanium dioxide; mesoporous; nanomaterials; environmental remediation; water remediation; NO$_x$; VOCs

---

1. Introduction

In recent years, one of the most important concerns of the scientific community and society has been health and environmental protection via a smart and sustainable use of natural resources. In this context, water resources are gaining increasing attention due to the occurrence of emerging pollutants including dyes, pharmaceutical and personal care products, endocrine disruptors, pathogens [1,2]. Moreover, the increasing amount of atmospheric pollutants has been regarded among the main causes of respiratory diseases such as emphysema, and bronchitis arising from the contact of NO$_x$ with lungs [3].

Unfortunately, conventional pollution remediation methods show limited performances. For instance, in the field of water treatment adsorption or coagulation methods aim at concentrating pollutants by transferring them to other phases; sedimentation, filtration, chemical and membrane technologies involve high operating costs and can generate toxic secondary pollutants in the ecosystem [4]; and chlorination, although widely used in disinfection processes, can generate by-products associated with cancer or other pathologies [5].

It turns out that the interest of the scientific community has been focusing on alternative methods such as the "advanced oxidation processes (AOPs)". AOPs are convenient innovative alternatives to conventional wastewater treatment processes [6,7] because they include a set of water treatment strategies such as ultraviolet (UV), UV-H_2O_2 and UV-O_3, and semiconductor-based photocatalysis that aim at accomplishing the complete mineralization of organic pollutants (i.e., their conversion into safe by-products such as O_2, H_2O, N_2 and mineral acids). Among AOPs, TiO_2-based photocatalysis has recently emerged as a promising water treatment [8]. Photocatalysis takes place upon the activation of a semiconductor with electromagnetic radiation from sun or artificial light. When exposed to electromagnetic radiation, a semiconductor absorbs photons with sufficient energy to inject electrons from the valence band (VB) to its conduction band (CB), generating electron hole pairs (e^-/h^+). The h^+ have an electrochemical potential sufficiently positive to generate $^\bullet$OH·radicals from water molecules adsorbed onto the semiconductor surface, while the e^- react with oxygen molecules to form the superoxide anions, $^\bullet O_2{}^-$, that quickly react with H^+ to finally produce $^\bullet$OH radicals after a series of concatenated reactions [9,10]. The overall photocatalytic efficiency depends on (i) the competition between e^-/h^+ recombination events and generation of reactive oxygen species (ROS) (ii) the competition between e^-/h^+ recombination events and e^-/h^+ trapping on semiconductor surface.

In this respect, TiO_2 nanoparticles (NPs) are extremely advantageous due to their high photoactivity, high chemical and photochemical stability, high oxidative efficiency, non-toxicity and low cost. In addition, the size-dependent band gap of nanosized semiconductors allows tuning the e^- and h^+ red-ox potentials to achieve selective photochemical reactions [11–13].

Remarkably, the reduced dimensions of TiO_2 NPs imply a high surface to volume ratio, which ensures a high amount of surface-active sites even upon immobilization of the photocatalyst onto substrates, thus avoiding the typical drop in performance due to the immobilization of bulk TiO_2. Immobilization is an essential requirement for a real application of TiO_2 NPs, both for safety and technological reasons [6]. Indeed, immobilization may limit accidental release of nanomaterials, thus preventing TiO_2 NPs turning into a secondary pollution source, and, at the same time, enables recovery and reuse of the photocatalyst. In fact, NPs have been demonstrated to harmfully impact on ecosystems, as reported in recent studies that have also shown that both TiO_2 NPs and TiO_2 NPs aggregates, at concentration higher than 10 mg/L, provoke hatching inhibition and malformations in the embryonic development of a model marine organism [14].

A great deal of work has been focused on improving the photoactivity of TiO_2 NPs and extending its optical response in the visible light range. Indeed, excellent reviews [15–19] and original papers [11,20–23] have overviewed the huge number of synthesis strategies aimed at purposely tailoring TiO_2 NPs by surface modification, doping, introduction of a co-catalyst, and crystalline structure manipulation.

Among the numerous strategies devoted to properly designing the morphological complexity of TiO_2 NPs, the possibility of obtaining mesoporous TiO_2 is attracting increasing interest [24]. The International Union of Pure and Applied Chemistry (IUPAC) classifies porous solids in three groups according to their pore diameter: namely microporous (diameter not exceeding 2 nm) and mesoporous (diameter in the range from 2 nm to 50 nm) and macroporous (diameter exceeding 50 nm) [25]. The porosity arises from the ordered or disordered assembly of individual nanocrystals (NCs) in larger structures (mesostructures). Ordered structures result from a regular arrangement of pores in the space and show a narrow pore size distribution, conversely disordered structures are characterized by a random aggregation of NPs, that gives rise to a large pore size distribution [23].

As a result, TiO_2-based mesoporous materials combine the well-known photocatalytic activity of TiO_2 with peculiar textural properties, including pore sizes and high specific surface areas, typical of NPs. Such features may contribute to increase the amount of absorbed organic pollutants and to dissolve the O_2 that can get to the TiO_2 surface thus improving the efficiency of the mineralization process [24]. Mesoporous TiO$_2$ NPs are regarded as promising adsorbents for various pollutants in water [26], as they present a high concentration of hydroxyl groups (−OH) on the surface, that allows adsorption of water pollutants and improves •OH radicals' generation, resulting in also being prone to further functionalization.

Moreover, TiO$_2$-based mesostructures and superstructures, such as hollow spheres, mesoporous TiO$_2$ nanotubes and mesoporous TiO$_2$ microspheres, enable multiple diffractions and reflections of incident UV light within the inner cavities, thus favoring a more efficient photogeneration of e$^-$/h$^+$ pairs, resulting in an improvement of the photocatalytic activity [27–29].

The present review aims at describing selected protocols, among the most interesting ones recently reported, for the synthesis of mesoporous TiO$_2$ with advantageous properties in terms of size/shape distribution, crystallinity and textural characteristics. Specifically, the presented synthesis protocols have been identified as suited to be implemented for a large-scale TiO$_2$ production, being scalable, cost-effective and relying on the use of safe chemicals. The high interest in the large scale manufacturing of nanoscale TiO$_2$ can clearly be seen when looking at the expectation of the complete conversion of TiO$_2$ production from bulk to nanomaterialt is foreseen to occur by 2025 with a production close to 2.5 million metric tons per year [30]. The review is mainly focused on sol-gel techniques and hydrothermal routes, namely soft templating approaches that make use of removable structure-directing agents as surfactant micelles, block copolymers, ionic liquids and biomacromolecules. All the reported protocols are suited for a viable scale-up because they make use of water as reaction solvent, and match the requirements of low-cost precursors, relatively low synthesis temperatures and high reaction yield.

Finally, an overview of the latest environmental applications of TiO$_2$ for water remediation and air purification will be presented.

2. Synthesis of Mesoporous TiO$_2$

2.1. Sol-Gel Methods

The sol-gel approaches [31] are among the most investigated techniques applied to obtaining ceramic or glass materials, having the advantages of being reproducible, industrially scalable and highly controllable.

The soft template processes underlying sol-gel strategies are generally based on several steps: (i) preparation of the solution of a selected TiO$_2$ precursor; (ii) hydrolysis of TiO$_2$ precursor in the presence of a suitable surfactant; (iii) removal of the solvent in order to facilitate the generation of the gel; (iv) condensation reaction; and (v) calcination for the complete removal of surfactant, solvent and unreacted precursor.

Among the sol-gel synthetic approaches the EISA (evaporation-induced self-assembly, Figure 1) has been recently applied for the preparation of metal oxides including TiO$_2$. The main feature of the EISA method is the use of a surfactant as a templating agent. Triblock copolymers as P123 (Poly(ethylene glycol)-block-poly(propylene glycol)-block-poly(ethylene glycol)) and F127 (poly(ethylene oxide) poly(propylene oxide)-poly(ethylene oxide)), are recognized as the most promising surfactants used for this method [32]. Indeed, surfactant selection represents one of the most critical parameters of EISA approaches because its chemical and physical properties affect the textural properties of the resulting material that can be deposited as a thin film on a suitable substrate.

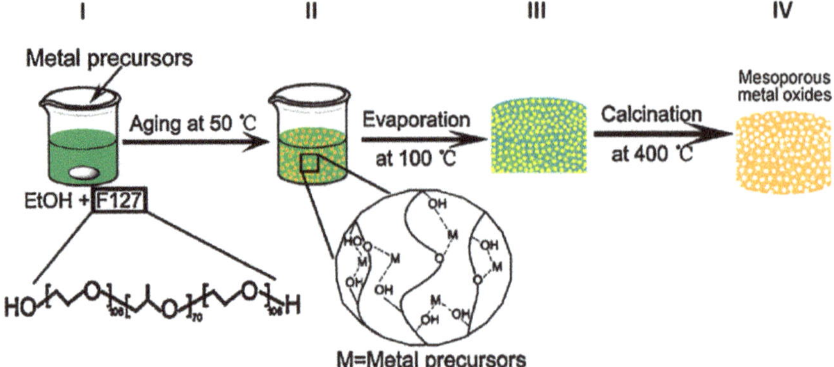

Figure 1. General synthetic scheme for the production of mesoporous metal oxides according to the evaporation-induced self-assembly method (EISA). The first step consists in the preparation of an ethanol solution containing the metal precursor (Ti(OBu)$_4$ for TiO$_2$) and the Pluronic F127 as templating agent (**I**). The mixture is kept at 50 °C for 24 h in order to induce the coordination bonds between the metal ions (M) and oxygen-containing group of F 127 (**II**). The subsequent thermal treatment at 100 °C for 6 h (**III**) promotes the formation of a xerogel of the metal-F127 hybrids. The final calcination at 400 °C (**IV**) is intended to remove of organic molecule and results in the formation of mesoporous metal oxides. Reprinted with the permission of ref. [33]. Copyright © 2019 4542370045937.

A typical sol-gel EISA synthesis of TiO$_2$ starts with the preparation of a solution containing Pluronic F127 in absolute alcohol (EtOH), and the subsequent addition of titanium butoxide Ti(OBu)$_4$ under vigorous stirring (Figure 1, I). The resulting suspension is kept at 50 °C for 24 h, and then dried at 100 °C for 6 h (Figure 1, II and III respectively). The as-prepared product shows a texture compatible with xerogels. The final calcination at 400 °C is carried out at specific heating rate in order to induce the removal the block copolymer surfactant species (Figure 1, IV). At this stage, aggregates formed by NPs of 5–10 nm in size have been produced, thus resulting in a mesoporous product with a specific surface area of 145.59 m^2/g and an average pore size of 9.16 nm [33].

M.G. Antoniou et al. reported a similar approach to obtain a mesoporous TiO$_2$-based coating for photocatalytic applications. The TiO$_2$ sol, comprised of titanium tetraisopropoxide (TTIP), acetic acid, isopropanol and Tween 80 as surfactant, is applied by dip-coating on glass substrate and then it is heated at 500 °C to remove the surfactant template. The dip-coating–calcination cycle is repeated 3 times for each deposition, resulting in uniform and transparent mesoporous nanocrystalline TiO$_2$ films with high surface area (147 m^2/g), porosity (46%) and anatase crystallite size of 9.2 nm. The amount of photocatalyst per cm^2 is estimated to be 62.2 mg/cm^2 with an overall coated area, considering both sides of the substrate, of 22.5 cm^2 [34].

An alternative strategy has been proposed to further increase the specific surface area of mesoporous TiO$_2$, that indicates the use of two types of TiO$_2$ precursors such as TiCl$_4$ and TTIP in a suitable molar ratio, with TiCl$_4$ playing the two-fold role of precursor and pH stabilizer. A solution containing a defined TiCl$_4$:TTIP:P123:ethanol ratio is stirred for 3 h at room temperature and the resulting product is suitable to be deposited by spin coating on glass substrates. After drying at room temperature for 24 h, the samples is thermally treated at 130 °C for 2 h to promote crossing-linking and prevent possible cracks in the film and collapsing of the mesostructure due to the high temperature. The final calcination treatment is carried out by heating stepwise up to 400 °C [35].

A recently reported sol-gel synthetic approach for the production of TiO$_2$ makes use of a biological template, namely the bacteriophage M13, a rod-shaped virus that is able to control the alkoxide condensation in the sol-gel process allowing the formation of mesopores having a diameter that can be tuned by adjusting only the reaction pH. Remarkably, the resulting product exhibits exceptional thermal stability of the anatase phase, which stays as the predominant phase even after a thermal treatment at 800 °C, that, in fact, promotes an increase in the pore and crystal size (Figure 2) [36].

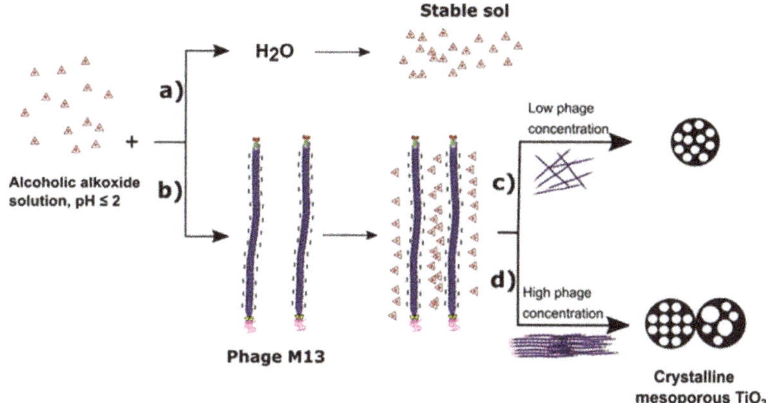

Figure 2. Proposed mechanism of mesoporous TiO$_2$ synthesis: consist in the preparation of a titanum alkoxide (titanium(tetra)isopropoxide) solution at pH ≤ 2. A vary stable sol is obtained with acid aqueous solution (pH 1–2) (**a**); the sol-gel reaction is performed with Phage M13 and a well-established structure is obtained (**b**). A local order of pores and macropores can be obtained at high phage concentration (**d**), while disordered pores with a narrow pore size distribution at a low concentration (**c**). Reproduced with permission from [36]. Copyright © 2019 4541961016973.

One of the main goals in the synthesis of mesoporous TiO$_2$ for environmental photocatalytic applications is to increase the TiO$_2$ optical response in the range of visible light. For this purpose, synthetic approaches have been developed to accomplish this result. For instance, a mixture of polyethylene glycol (PEG) and polyacrylamide (PAM), has been used as the templating agent. PAM and PEG are slowly introduced in a mixture of deionized water, nitric acid (8%), ethanol and Ti(OBu)$_4$ as TiO$_2$ precursor. The resulting white gel is dried until a light-yellow powder is obtained that undergoes two calcination steps: the first in nitrogen atmosphere, and the second in air. In the first calcination step, three different temperature values are investigated: 500 °C, 600 °C and 700 °C respectively, while the second calcination step is carried out at 500 °C. The authors have demonstrated how increasing PAM mass the gel formation rate increases, due to the improved interaction between amide groups of PAM with the hydroxyl groups of the TiO$_2$ sol. The PEG prevents the mesostructure collapsing during the first thermal treatment. Moreover, the molecular weight (MW) of PEG has been reported to increase as the crystallite size increases and the specific surface area decreases. The obtained mesoporous TiO$_2$ is found to have a specific surface area measured by BET (Brunauer–Emmett–Teller) test between 104.25 and 110.73 m^2/g and a pore size (measured by Barret-Joyner-Halenda isotherm) between 16.92–16.80 nm; being the variation of the specific surface area and pore size values affected by the variation of the molecular weight of the PEG used in the synthesis. The authors point out that the two calcination steps improve the textural properties of the TiO$_2$ because they promote a higher crystallinity, and allow to achieve a homogenous porosity, and a higher specific surface area. In particular, the first calcination step under N$_2$ atmosphere causes the conversion of PEG (less thermally stable than PAM) in amorphous carbon, which plays the role of a scaffold around pores, thus preventing the mesostructure from collapsing [37]. Furthermore, the small amount of amorphous carbon is able to induce a doping effect, and therefore the obtained photocatalyst is able to extend its photoactivity to visible range, as demonstrated in the ultraviolet–visible (UV–Vis) reflectance spectrum, that shows an increase, in the visible range, of the Kubelka–Much function intensity.

Also, Phattepur et al. have synthesized mesoporous TiO$_2$ with an innovative sol-gel technique by using lauryl lactyl lactate as biodegradable and inexpensive additive to control the size of large inorganic cluster. The nanostructured photocatalyst is prepared by using Ti(OBu)$_4$ as precursor in a solution containing a defined amount of lauryl lactyl lactate (0.25 mL, 0.5 mL, 0.75 mL and 1 mL), ethanol, and hydro-chloric acid. Such a solution is mixed with a second solution of ethanol and distilled water under vigorous stirring for 8 h up to the generation of the gel. The mixture is, then, aged, dried, ground and finally calcined, resulting in a final product with a specific surface area up to 40.10 m^2/g, a pore volume 0.112 cm^3/g [38].

The interest towards colloidal routes for the synthesis of mesoporous TiO$_2$ is further supported by a recently granted US patent [39]. The synthetic scheme consists of an acid-catalyzed hydrolysis of a water soluble TiO$_2$ precursor as TiOCl$_2$ or TiOSO$_4$, occurring in the presence of a porogen molecule, namely an organic alpha hydroxyl carboxylic acid, as citric acid, at relatively low temperatures (up to 100 °C). Such a procedure allows 100 nm spherical mesoporous anatase NPs to be achieved and control of the pore size by varying the molar ratio between the TiO$_2$ precursor and the organic alpha hydroxyl acid. Remarkably, these mesoporous TiO$_2$ NPs show a bimodal pore size distribution. Such bimodal porosity is due to the presence of both intra-particle and inter-particle pores. Intra-particle pores (from 2 nm to 12 nm in size) are detected in individual TiO$_2$ NPs while the inter-spatial fissures of dimensions from 15 nm to 80 nm, due to the packed arrangement of the NPs, result in inter-particle pores [39].

2.2. Synthesis in Room Temperature Ionic Liquids

Room-temperature ionic liquids (RTILs) are regarded with increasing interest both in academia and industry [40]. In the synthesis of TiO$_2$ NCs, the use of RTILs offers multifold advantages including the control over the morphology and phase composition, the colloidal stability [41,42] and the possibility to achieve a large scale production of photocatalytic TiO$_2$ [41,43,44]. Indeed RTILs allow, in principle, to perform synthesis of titania at low temperature as they are organic salts characterized by a low melting point (lower than 100 °C), thus resulting in being liquid and thermally stable over a wide temperature range [45]. In a typical synthesis of TiO$_2$ NPs, the 1-butyl-3-methylimidazoliumtetrafluoroborate (BF) is used as solvent and TiCl$_4$ as TiO$_2$ precursor. After mixing BF and TiCl$_4$, purified water is added slowly under vigorous stirring at room temperature to promote the immediate hydrolysis of TiCl$_4$ indicated by the appearance of turbidity. Such a turbid solution is stirred at 80 °C for another 12 h and the resulting product is collected by centrifugation upon dilution with water in order to decrease the viscosity due to the BF. The residual solvent is removed by extraction with acetonitrile, soluble to both the inorganic species and the RTIL, in a closed vessel at 50 °C for 8 h and the final product is dried in a vacuum oven at 40 °C [43]. Such a synthetic approach is used to synthesize mesoporous N-doped TiO$_2$ (10–50 nm in size) in anatase phase, with 5–8 nm pores observed by transmission electron microscopy (TEM) and a band gap, determined by diffuse reflectance spectroscopy, of 2.47 eV [46].

2.3. Hydrothermal Synthetic Methods

A typical hydrothermal process is carried out in an autoclave, possibly equipped with Teflon liners and a closed system under controlled temperature and/or pressure (room temperature and at pressure >1 atm) [47]. The temperature can be higher than the boiling point of water corresponding to the pressure of vapor saturation [48].

Hydrothermal processes are extremely attractive for the large-scale production of mesoporous TiO$_2$ NPs, because they (i) are environmentally friendly, (ii) make use of aqueous solutions, (iii) do not require any post-calcination treatment and (iv) allow a facile recovery of the photocatalyst after the synthesis [48–50].

TiO$_2$ NPs prepared by hydrothermal methods show several advantages, including high crystallinity, reduced particle size, uniform size distribution, prompt dispersibility in polar and non-polar solvents and a stronger interfacial adsorption; moreover, they enable the easy fabrication of high-quality coatings on several supporting material [51].

In hydrothermal synthesis, temperature, filling volume [52] pressure [53] pH and treatment duration are regarded as the key parameters to control the resulting morphological and structural properties of TiO_2. Under hydrothermal conditions, a decrease of the reaction temperature cause a particle size decrease and an increase of particle agglomeration [47]. The growth of TiO_2 NPs is also possible by using a template-based technique, taking advantage of the use of suitable high-molecular-weight surfactants, able to promote structural self-assembly [47].

The hydrothermal method applied on TiO_2–nH_2O amorphous gel, considered as TiO_2 NPs precursor, has been widely used to prepare nanocrystalline titania [54]. This treatment can be carried out either in pure distilled water or in the presence of mineralizing species, such as hydroxides, chlorides and fluorides of alkali metals at different pH values [54]. Kolen'ko et al. have developed the synthesis of ultrafine mesoporous titania powders in anatase phase via hydrothermal process starting from TiO_2–nH_2O amorphous gel. The preparation of TiO_2–nH_2O amorphous gel requires multiple steps starting from the high-temperature hydrolysis of complex titanyl oxalate acid ($H_2TiO(C_2O_4)_2$) aqueous solutions. Briefly, the preparation route of $H_2TiO(C_2O_4)_2$ aqueous solution is based on: (I) the preparation of H_2TiCl_6 from $TiCl_4$ and chloride acid (HCl), (II) hydrolysis of H_2TiCl_6, (III) wash of TiO_2–nH_2O by distilled water, and (IV) dissolution of TiO_2–nH_2O in the oxalic acid. At this stage the TiO_2–nH_2O is treated in a polytetrafluoroethylene (PTFE)-lined autoclave at temperature of 150 °C or 250 °C for a period of time ranging from 10 min to 6 h. After the treatment in autoclave the sample is cooled down to room temperature, the obtained product is centrifuged, washed and dried at 80 °C [54]. Washing procedures are essential in order to collect the photocatalyst and remove the templating agent and salts that can possibly obstruct the pores, thus achieving the desired porosity and specific surface area [55,56]. The size of the mesoporous anatase particles can be controlled in the range of 60–100 nm by adjusting the concentration of $H_2TiO(C_2O_4)_2$ aqueous solution that is expected to affect the amount of TiO_2 nuclei.

Hydrothermal synthesis allows also a fine control over NP morphology. Indeed, mesoporous TiO_2 microparticles with a well-defined spherical shape in the range of 2–3 µm, and a crystallite size in the range from 7.3 nm to 22.3 nm have been prepared by hydrothermal reaction of poly(ethylene glycol)-poly(propylene glycol)-based triblock copolymer and TTIP mixed with 2,4-pentanedione [57]. The surfactant solution is prepared dissolving the triblock copolymer in distilled water at 40 °C and adding sulfuric acid. Successively, TTIP is mixed with 2,4-pentanedione and slowly added dropwise into the previously prepared surfactant solution. The reaction is carried out at 55 °C for 2 h without stirring and a light-yellow powder is obtained. The hydrothermal treatment is performed at 90 °C for 10 h followed by an annealing step, which is necessary to remove the residual surfactant.

Zhou and co-workers have proposed titanium sulfate ($Ti(SO_4)_2$) as a precursor of TiO_2 in the presence of urea to obtain microspheres by hydrothermal treatment TiO_2 [58], by using reaction time as a key parameter to control average crystallite size, pore size and volume, and the specific surface area.

Mesoporous TiO_2 anatase microspheres have been also successfully obtained with a simple one-step hydrothermal synthesis by Lee et al. [49]. The titanium-peroxo complex is treated with nitric acid, 2-propanol, NH_4OH in Teflon-lined autoclave at 120 °C for 6 h. The final product is filtered, washed with distilled water several times until the pH reaches 7 and dried at 65 °C for 1 day, obtaining the mesoporous material without any post-calcination procedure. In particular, the size of mesoporous TiO_2 anatase microspheres lays in a dimensional range between 0.5 µm and 1 µm. According to high-resolution TEM (HRTEM) analysis, the formation of a microsphere (secondary particle) results from the aggregation of diamond-shaped TiO_2 NCs (50 nm × 20 nm) (primary NPs). The authors have suggested that during the hydrothermal reaction the aggregation of individual diamond-shaped TiO_2 NCs in secondary particles, is thermodynamically favored with respect to the formation of primary TiO_2 NPs. Interestingly, the porosity of the microspheres is ascribable to the interspaces between TiO_2 NPs assembled to form a microsphere [49].

Similarly, Santhosh et al. have synthesized mesoporous TiO_2 microspheres by a facile hydrothermal reaction carried out at 120 °C for 24 h that, instead, makes use $Ti(OBu)_4$ as precursor [59]. The material structure is based on TiO_2 having a diameter in a range from 100 nm to 300 nm. The corresponding specific surface area is 56.32 m^2/g and the bimodal pore structure shows pore width of 7.1 nm and 9.3 nm respectively [59].

Hollow TiO$_2$-based core-shell structures have been also reported and they are expected to show a high photocatalytic activity because their unique morphology allows the multiple reflection of UV light within the inner cavity [60]. Furthermore, they display as valuable advantages high surface-to-volume ratio, low density and low production cost. Recently, Cui et al. have reported a facile one-step hydrothermal method to synthetize a mesoporous hollow core-shell structured TiO$_2$ microspheres. PEG (MW 2000) has been used as soft templating agent to obtain Ti^{4+}-PEG globules suited for the generation of hollow structures. The control on crystallite size, microsphere size, shell thickness and the roughness are achieved by tuning the duration of the hydrothermal treatment. In particular, increasing the reaction time, the crystallite and microsphere size, the shell thickness and the roughness increase while the core size decreases [60].

Ye et al. have developed an interesting strategy for the large–scale synthesis of hallow microspheres by performing a hydrothermal treatment followed by a calcination step (Figure 3) [61]. They have used potassium titanium oxalate (PTO) as Ti precursor performing the hydrothermal process in autoclave at 150 °C for 4 h, thus obtaining 1.75 μm microspheres composed of TiO$_2$ NPs. In particular, a strong effect of calcination temperature on the photocatalytic performance of the photocatalyst is reported, with the hollow TiO$_2$ microspheres calcined at 500 °C showing the highest photocatalytic activity.

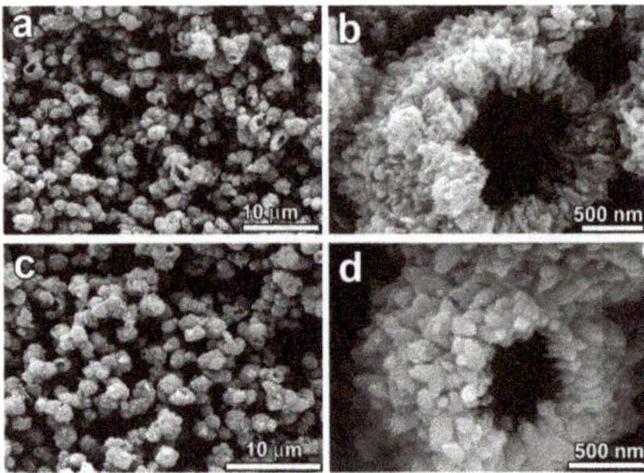

Figure 3. Large–scale synthesis of TiO$_2$ hallow microspheres has been obtained by Ye et al. by using a hydrothermal treatment following the calcination. The morphology and microstructure of the hollow microspheres have been characterized by scanning electron microscopy (SEM) (**a**–**d**), in order to evaluate the effect of the calcination. (**a**,**b**) SEM micrographs of the not calcined sample: low-magnification and high-magnification, respectively. (**c**,**d**) SEM micrographs of the TiO$_2$ hollow microspheres calcined at 500 °C for 2 h: low-magnification and high-magnification, respectively. SEM micrographs of the non-calcined material show that the sample consisting of large-scale hollow microspheres, being the shell of the microsphere itself made of numerous nanoparticles (NPs). Interestingly, the SEM micrographs of calcined material indicate an excellent thermal stability of the hallow microsphere. The morphology and the mean external diameter of the hollow microspheres has been found unchanged after the calcination treatment at 500 °C. Reproduced with permission from [61]. Copyright © 2018 Elsevier.

TiO$_2$ hollow spheres aggregates, characterized by porous walls, can be also synthesized on large-scale by performing the hydrothermal hydrolysis of Ti(SO$_4$)$_2$ assisted by NH$_4$F without employing any templates [62]. The resulting hollow material, reported in Figure 4, presents a high surface area, smaller crystal size, and highly porous structure [62] Several hydrothermal protocols have been also successfully developed to synthesize anisotropic mesoporous NPs. Anisotropic TiO$_2$ NCs have been extensively investigated for several energy conversion applications, since their peculiar morphology enables a facile charge transport along the longitudinal dimension and decreases the e$^-$/h$^+$ recombination rate [51,52].

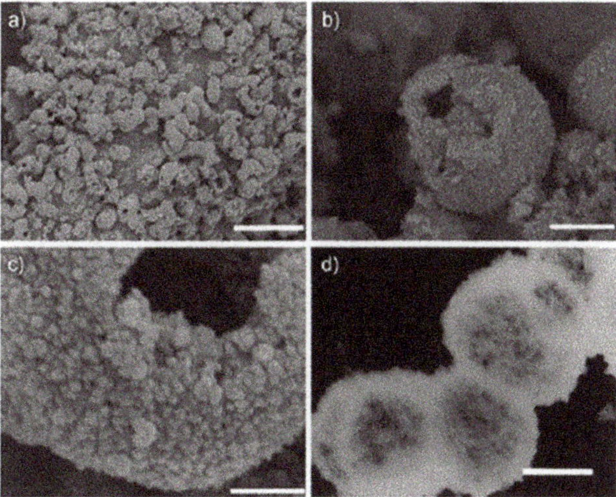

Figure 4. Large-scale synthesis of porous TiO$_2$ hollow aggregates reported by Liu et al. The synthesis is carried out using a low-temperature hydrothermal method without templates. (**a**–**c**) Field-emission scanning electron microscopy (FESEM) analyses at different magnifications and (**d**) transmission electron microscope (TEM) micrograph of the "as prepared" sample by one-step hydrothermal treatment at 160 °C for 6 h. The scale bars for (**a**–**d**) are 5 mm, 500 nm, 100 nm, and 500 nm, respectively. The low-magnification FESEM micrograph (**a**) of the sample indicates that the aggregates are composed of a large amount of NPs. While, the micrograph (**b**) shows the hollow interior of the single aggregate and the micrograph (**c**) at a higher magnification highlights the porous structure of the TiO$_2$ NP aggregates. Moreover, the hollow structure of the TiO$_2$ sample is confirmed by TEM micrograph (**d**). Reproduced with permission from [62]. Copyright © 2018 John Wiley and Sons.

Mesoporous TiO$_2$ nanotubes (TNTs) are materials of great interest due their high ion-exchange capability, relative stability, enhanced conductivity and high specific surface area [63,64]. In addition, TNTs contain a large amount of hydroxyl groups respect to spherical TiO$_2$ and may find an effective use as ion adsorbent systems. Sattarfard et al. have prepared mesoporous TNTs by hydrothermal synthesis starting from TiO$_2$–P25 NPs as titania precursor [63].

In this protocol, TiO$_2$–P25 powder is introduced into NaOH aqueous solution and held at 115 °C for 24 h in a stainless-steel with Teflon lining reactor placed in an oil bath. Afterwards, the mixture is cooled to room temperature and centrifuged, and the obtained precipitate requires several washing, recovery and thermal treatment steps to finally obtain TNT with a specific surface area of 200.38 m^2/g and an average outer and inner diameter approximately of 9 nm, 4 nm, respectively, with a wall thickness of 2.5 nm. Remarkably, the use of a strong aqueous base as NaOH as dispersing agent for TiO$_2$ P25 and the subsequent hydrothermal treatment are the key steps that determine the formation of titanate tubular structures, while washing with HCl has been demonstrated to be essential to convert tubular titanate in TiO$_2$ nanotubes. Indeed, the treatment of TiO$_2$ NPs with NaOH is known to break Ti–O–Ti bonds, thus resulting in the formation of sheets, while the subsequent washing with acid or water, reducing the electrostatic charge, induces the folding of the sheets yielding the formation of nanotubes [63].

TiO$_2$ nanowires can be also successfully obtained by using hydrothermal method, resulting in a fine shape control, as proposed by Asiah et al. that investigated a low-cost, high-purity shape-controlled synthetic strategy for the large-scale production of mesoporous TiO$_2$ nanowires [65]. Titanium (IV) oxide nanopowder is used as Ti precursor solution and treated with NaOH aqueous solution. The treatment is performed by following a hydrothermal route in autoclave in a Teflon beaker at 150 °C for a reaction time ranging from 1 to 10 h. The obtained white precipitate is washed with HCl and deionized water until pH = 7 is reached. The final product, dried at 40 °C overnight and annealed at 500 °C, is found to consist of few microns long nanowires with diameters from 15 to 35 nm, according to the duration of the treatment (Figure 5). Interestingly, the hydrothermal method has also been shown to be powerful for synthesizing a three-dimensional (3D) TiO_2 mesoporous superstructure [66,67]. TiO_2 based superstructures with a branched architecture present an enhanced specific surface area, improved charge separation and transfer within the TiO_2 branches, thus increasing the e^-/h^+ pairs lifetime and, therefore, the ROS generation [68].

Figure 5. Large-scale hydrothermal synthesis is explored to produce surfactant-free seed mediated mesoporous TiO_2 nanowires by Asiah et al. (**a**–**h**) FESEM micrographs of TiO_2 nanowires at different growth time, (**a**) 1 h, (**b**) 2 h, (**c**) 3 h, (**d**) 4 h, (**e**) 5 h, (**f**) 6 h, (**g**) 8 h and (**h**) 10 h, respectively. (**i**) Schematic diagrams of growth evolution of TiO_2 nanowires in time. The effect of hydrothermal growth time on the evolution of the morphology and structural properties of mesoporous TiO_2 nanowires is shown. The initial morphology of as-prepared TiO_2 reacted for 1 h (**a**) consists of unreacted NPs which are agglomerated and small yield of the formed nanowires with very low aspect-ratio. After 2 h (**b**) the structure of nanowires is clearly formed. However, the NPs are completely converted into nanowires after 3 h (**c**). Reproduced with permission from [65]. Copyright © 2018 Elsevier.

Baloyi et al. have proposed the hydrothermal synthesis of dandelion-like structures from $TiCl_4$ and water via a simple hydrothermal [68] synthesis realized in a reaction flask by adding drop-wise $TiCl_4$ to the super-cooled high purity water, by using a separator funnel under vigorous stir and heating at 100 °C for 24 h. Successively, the suspension is centrifuged and washed with water to remove any chloride ions from the solid TiO_2 and the resulting solid is dried for 16 h overnight at 120 °C. It has been proposed the obtained nanostructures grow according to a four-step reaction mechanism: (i) nucleation and NP formation; (ii) formation of spheres; (iii) further growth; and (iv) formation of flower-like TiO_2 structures by agglomeration of the dandelions.

Pan et al. have reported the large–scale synthesis of uniform urchin-like mesoporous TiO_2 hollow spheres (UMTHS) by a hydrothermal method based on targeted etching of self-organized amorphous hydrous TiO_2 solid spheres (AHTSSs) (Figure 6) [67]. The growth of the UMTHSs under hydrothermal conditions has been proposed to start from the spontaneous reconstruction of surface–fluorinated AHTSSs in the presence of surface coating of polyvinylpyrrolidone (PVP). Briefly, the previously synthesized AHTSSs are treated with NaF, used as etching agent and successively with PVP. After an hour of stirring, the suspension is transferred to a Teflon-lined autoclave and kept at 110 °C for 4 h. The UMTHSs are obtained by collecting, washing with diluted NaOH solution and water, and finally calcining at 350 °C for 2 h.

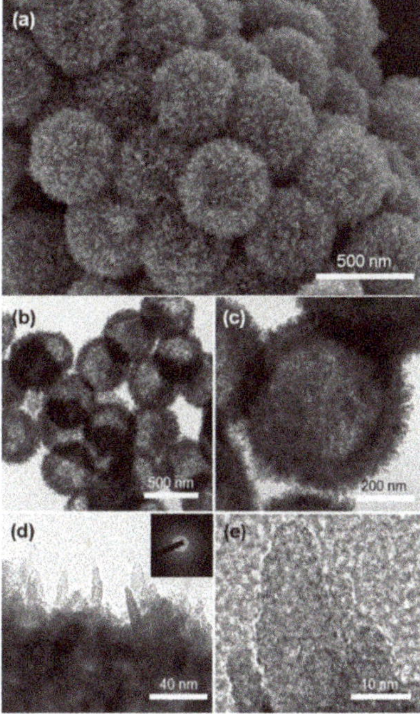

Figure 6. Large-scale synthesis of urchin-like mesoporous TiO_2 hollow spheres (UMTHS) by low-temperature hydrothermal method. SEM (**a**) and TEM (**b–e**) micrographs of UMTHS obtained by calcining the powder after hydrothermal reaction. SEM micrographs shows that the UMTHS are monodisperse with uniform particle size and consist of radially arranged anatase nanothorns, assembling an urchin-like shaped hierarchical structure. Furthermore, TEM analysis (**b**,**c**) indicates that the spherical shell consisting of radial nanothorns. The single hallow sphere is polycrystalline due to radial orientation of nanothorns, as revealed by selected-area electron diffraction (inset of **d**). Moreover, the high-resolution TEM (**e**) shows that each nanothorn presents a single-crystal nature and possesses the lattice fringes of anatase (101) plane with a d -spacing of 0.35 nm aligned over the single nanothorn. Reproduced with permission from [67]. Copyright © 2018 John Wiley and Sons.

The inner structures of the materials have been demonstrated to be tunable from the conventional solid spheres to hollow and complex core-shell and yolk-shell configuration, by varying the experimental parameters and suitably protecting the inner structure by prefilling AHTSSs with PEG. The synthesized structures show a large surface area up to 128.6 m^2/g and excellent photocatalytic performances for environmental application [67].

Recently, Hu et al. have employed one-step hydrothermal method to synthetized 3D flower-like TiO$_2$ microspheres using Ti(OBu)$_4$ as titanium source and glacial acetic acid (HAc) as solvent and capping agent, at the same time [69]. Briefly, a solution of Ti(OBu)$_4$ and HAc is prepared and, after stirring, is transferred into a Teflon-lined stainless-steel autoclave and heated at 140 °C for a defined period. The resulting product is collected upon centrifugation, washed with ethanol and deionized water repeatedly, dried at 60 °C for 12 h, and annealed in air. The 3D flower-like structures are formed due to the oriented assembly of nanosheets as reported in Figure 7.

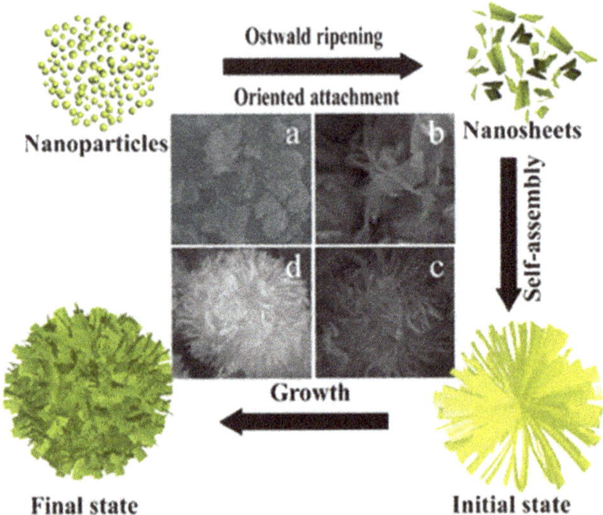

Figure 7. Schematic illustration of the growth evolution mechanism for 3D flower-like TiO$_2$ microspheres self-assembled by nanoplates using the hydrothermal method at different time: (**a**) 3 h; (**b**) 6 h; (**c**) 9 h; (**d**) 12 h. TiO$_2$ NPs after the nucleation begins to grow in the direction of orientation, forming nanosheets after dehydration. With the increasing of hydrothermal time (at 400 °C), the nanosheets are self-assembled to form a 3D flower-like structure as final state. Reproduced with permission from [69]. Copyright © 2018 Springer Nature.

TiO$_2$ based photocatalyst with wormhole-like disordered structure was hydrothermally prepared by using titanium sulfate (Ti(SO$_4$)$_2$) as precursor of TiO$_2$, and cetyltrimethyl ammonium bromide (CTAB) as a structure-directing agent [70]. The resulting photocatalyst is characterized by a narrow pore size distribution and a very high surface area of 161.2 m^2/g, presenting an excellent adsorption ability for the removal of methyl orange (MO) and Cr(VI) from wastewater. The Ti(SO$_4$)$_2$ has been used as TiO$_2$ precursor also for the hydrothermal synthesis of hollow TiO$_2$ microspheres composed of 60 nm sized nanospheres thus resulting in a large hierarchical nanostructure characterized by a surface area and pore volume of 123 m^2/g and 0.19 cm^3/g, respectively [71]. The main characteristics of the synthetic approaches reported in this section have been summarized in Table 1.

Table 1. Main characteristics of the reported synthetic approaches for the preparation of mesoporous TiO_2.

Synthetic Routes	Control Parameters	Particle Size	Pore Size	Specific Surface Area	Main Advantages	Main Drawbacks
Sol-gel methods [33–39]	pH, calcination temperature	Micrometer and sub-micrometer aggregates of nanoparticles	Pore volume from 0.18 cm^3/g to 0.50 cm^3/g; Pore size from 9.16 nm to 16.9 nm	From 70 m^2/g to 150 m^2/g	Low cost; User-friendly protocol Special facilities not required	Difficult control on the morphological and textural properties; calcination step required
Synthesis in room temperature ionic liquids [41–46]	Viscosity, temperature, stirring	Crystallite size 3–6 nm	Pore size 5 nm to 8 nm	554 m^2/g	Low temperature required	High cost due to the use of ionic liquid
Hydrothermal synthesis [47–71]	pH, temperature, pressure, filling volume, hydrothermal treatment duration	Micrometric particles formed by nanoparticles from 3.4 nm to 27 nm	Pore volume from 0.18 cm^3/g to 0.50 cm^3/g Pore size from 3.4 nm to 27 nm	From 25 m^2/g to 395 m^2/g	High control over morphology and textural properties; Use of aqueous suspensions; Calcination step not always required: Facile photocatalyst recovering	Use of a specific facility (autoclave)

3. Environmental Applications of Mesoporous TiO$_2$

3.1. Application for Water Treatment

Several studies have demonstrated the effectiveness of TiO$_2$ assisted photocatalysis in the removal of contaminant of emerging concern including, pesticides, synthetic and natural hormones, industrial chemicals pharmaceuticals and personal care products (PPCPs). PPCPs represent a relevant example because they are continuously released into the aquatic environment as they are extensively used in human and veterinary medicine. However, they cannot be removed using conventional wastewater treatments, thus representing a potential risk to aquatic organisms and public health. Despite mesoporous TiO$_2$ is considered extremely promising for photocatalytic treatment of water, the investigations of its photocatalytic properties have been performed mainly on model molecules as organic dyes. The present section, instead, intends to report on the photocatalytic studies, currently presented in literature, on the application of mesoporous TiO$_2$ for the photocatalytic gradation of molecules of environmental concern.

The photocatalytic activity of 25–30 nm mesoporous TiO$_2$ NPs with a surface area up to 40.10 m^2/g, prepared by the sol-gel protocol described in [38] has been investigated against degradation of salicylic acid, caffeine and phenol. Phenol and its derivatives are major water pollutants in the chemical, textile, petrochemical, and paint industries. They are carcinogenic, mutagenic, and phenol derivatives can jeopardize mammalian and aquatic life [72]. Wastewaters coming from the cosmetic, paper mill, human, veterinary drugs and pharmaceutical industries, often contain salicylic acid and caffeine, which is a hazardous substance for human health [73,74]. The photocatalytic experiments have been performed using the mesoporous TiO$_2$ in water suspension and uner UV irradiation with encouraging results in terms of degradation performance. Indeed, after 4h of UV light irradiation, a degradation efficiency of 92% and 59% has been achieved for phenol and caffeine respectively, while for the salicylic acid a degradation amount of 88% has been reached after 3 h of reaction [38].

Hollow of TiO$_2$ microspheres have demonstrated to be effective in the photocatalytic degradation of 4-chlorophenol (4-CP), promoting the 90% photodegradation of 4-CP within 1 h under UV light irradiation (Figure 8). The potential reusability of TiO$_2$ hollow microspheres, has been investigated by performing three cycles of reuse and any significant change of the photocatalytic performance has been detected. Also, no relevant alteration in terms of the 4-CP degradation rate constant has been measured, thus highlighting the effectiveness of this material for photocatalytic water treatment [75].

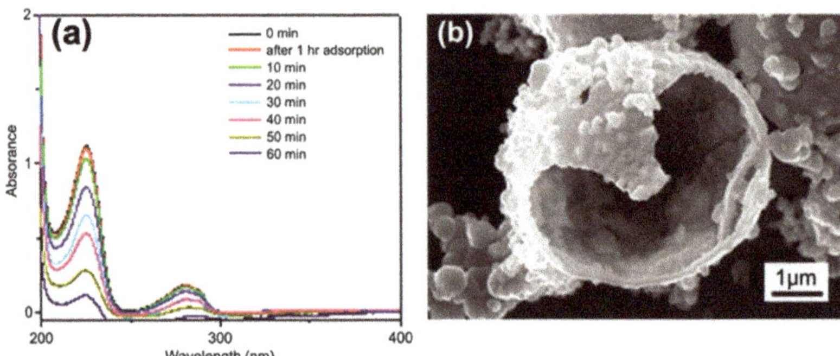

Figure 8. (**a**) Ultraviolet (UV)–visible absorption spectra of the 4-CP containing sample, in the course of the heterogeneous photocatalytic experiment assisted by hollow TiO$_2$ microspheres as photocatalyst. The TiO$_2$ has been separated from the aqueous solution of 4-CP by filtration, before recording the absorption spectrum (**b**) FESEM high magnifications micrography of TiO$_2$ hollow microspheres. Reproduced with permission from ref. [75] http://creativecommons.org/licenses/by-nc-nd/2.5/in/.

3.2. Air Treatment

Photocatalytic processes, assisted by TiO_2, have in recent years shown a great potential in the abatement air pollutants as NO_x gases (commonly referred to as nitrogen monoxide, NO, and nitrogen dioxide NO_2), and volatile organic compounds (VOCs) [76]. VOCs represent a major group of indoor air pollutants responsible of the production of tropospheric ozone and secondary organic aerosol with several adverse health effects [77]. Indeed, several VOCs such as halogenated hydrocarbons, ketones, alcohols and aromatic compounds are toxic and carcinogenic pollutants. Moreover, the interaction of NO_x with VOCs provokes the formation of by-products even more dangerous than NO_x gases, such as nitrous acid and peroxyacyl nitrate (PAN) [3]. Recently, several examples concerning the application of mesoporous TiO_2 material to removal of NO_x and VOCs have been reported in literature [78,79].

3.2.1. Photocatalytic Abatement of NO_x by Using Mesoporous TiO_2

Mesoporous TiO_2-based films are widely studied for photodecomposition processes based on the surface-adsorbed reactants, because the large surface area of mesoporous TiO_2 determines an increased amount of adsorbed resulting in an enhancement of their decomposition rate [79,80]. A relevant example of mesoporous TiO_2-based film applied to air purification has been reported by Kalousek et al. The TiO_2-based film is synthetized by a template-assisted method based on the evaporation-induced self-assembly mechanism which allows to control the film thickness and consequently its surface area and pore volume [79]. NO_x degradation assisted by porous films has been investigated and compared with that achieved by means of commercial Pilkington Active Glass. The film obtained exhibits a very high photocatalytic efficiency probably due to their peculiar morphological properties such as large surface area and pore volume. Such a valuable photoactivity has been assumed to arise from the local increase of compound's partial pressure reacting in the nanopores or cavities of the film in proximity to the photocatalytic sites. Moreover, the high selectivity towards nitric acid has been ascribed to the strong adsorption of the intermediate products. Balbuena et al. have reported as innovative solution for environmental remediation, mesocrystalline anatase NPs synthesized by hydrothermal method easily up scaled to industrial level [81]. The obtained TiO_2 photocatalyst (namely TiO_2 A350) exhibits a suitable nanostructure and porosity, with a surface area and pore volume of 63.5 m^2/g and 0.22 cm^3/g, respectively. The oxidation of NO is evaluated by using a laminar flow reactor and an artificial sunlight as irradiation (25 and 550 W m^{-2} for UV and visible irradiances, respectively). The final product presents a higher degradation efficiency and selectivity for the NO_x abatement than TiO_2 P25 (Figure 9), due to probably to the presence of mesopores which increases the surface area of the samples and the accessibility for the reactant molecules.

3.2.2. Photocatalytic Abatement of Volatile Organic Compounds (VOCs) by Mesoporous TiO_2

Similarly, in NO_x degradation, photocatalytic processes assisted by TiO_2 surfaces may potentially remove VOCs, because photogenerated $^\bullet$OH radicals and superoxide radicals can promote their mineralization leading to water vapour H_2O and carbon dioxide (CO_2) [15,82]. Generally, the degradation of VOCs leaves a recalcitrant carbonaceous residue accumulation on the photocatalyst surface as a result of incomplete oxidation of VOCs, which causes catalyst deactivation decreasing the photocatalytic efficiency [83]. Several efforts have been devoted to remove this drawback that, in spite of the long-term photocatalytic stability, still represents a big challenge for the degradation process of VOCs. A possible solution is given by the use of mesoporous TiO_2, that, due to the channels in the structure, increases the density of highly accessible active sites thus promoting the diffusion of reactants and products as a consequence of the capability of absorbing pollutant molecules [83]. A mesoporous TiO_2-based system has been fabricated by Ji et al., as a promising photocatalyst for the oxidation of gaseous benzene [83]. This material exhibits a higher photocatalytic performance and stability towards benzene degradation compared to commercially available TiO_2 P25. In particular, among others, the material obtained upon calcination at 400 °C has shown a good mesoporous structure and superior capacity for benzene adsorption. Nowadays, photocatalytic materials for filter represent another relevant application devoted to air purification systems [84]. In order to fabricate effective photocatalytic filters porous foams have been identified as solid supports due to their high open porous structure with large surface area, that provide excellent structure for gas passing while maintaining a high level of surface contact and low level of pressure drop. Among the photocatalytic filters for commercial environmental purifiers, TiO_2-immobilized ceramic foams represent a very effective system, especially for reducing indoor air pollution caused by VOCs [84]. Recently, Qian et al. have fabricated hydro-carbon foams (CFs) as support for TiO_2 photocatalyst for an indoor air treatment application resulting in TiO_2/hydro-CFs [82]. These CFs are prepared by waste polyurethane foams (PUF), while phenolic resin is used as a hard template and a carbon source. Mesoporous TiO_2 coatings are uniformly deposited on the foam by a self-assembly sol-gel method. The TiO_2/hydro-CF exhibit obtained enhanced photocatalytic oxidation activity for VOCs under UV–vis and visible light irradiation, probably due to the synergic combination of TiO_2 and CF that promote the mass transfer and the accumulation of pollutant molecules at the interface photocatalyst/substrate according to the mechanism proposed in Figure 10.

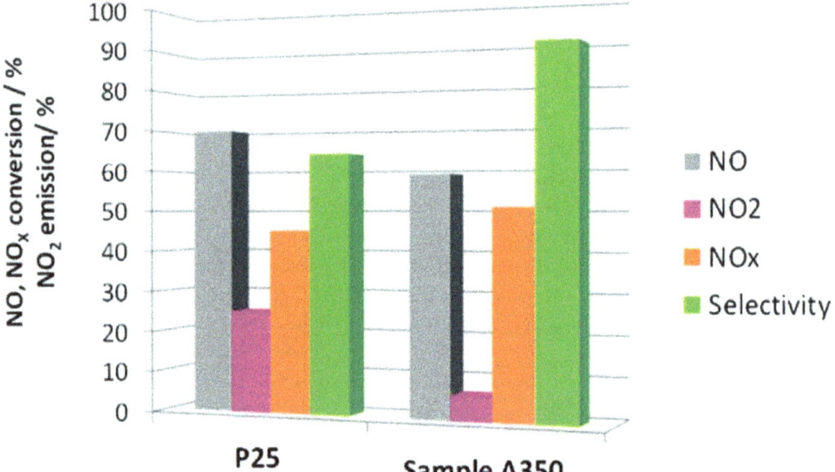

Figure 9. Mesocrystalline TiO_2 NPs are synthesized using a hydrothermal method as a photocatalyst for NOx abatement. The diagram reports NO conversion (%, grey), NO_2 released (%, pink), NOx conversion (%, orange) and selectivity values (%, green) for A350 Titania ("as prepared" sample) and P25 TiO_2 (as reference material) after five hours of light irradiation. The selectivity is here intended as the complete conversion of NO in nitrate or nitric acid, that leads to the efficient removal of NOx species. In terms of NOx removal capability, the prepared titania appears to be a desirable photocatalyst, which combines high efficiency in photochemical NO conversion (grey bar) and selectivity (green bar), resulting in the highest NOx conversion (orange bar). Reproduced with permission from [81]. Copyright © 2019.

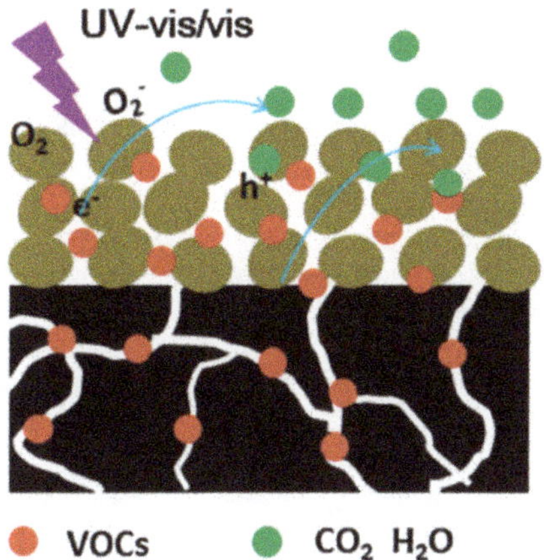

Figure 10. Schematic mechanism of the proposed photocatalytic oxidation of volatile organic compounds (VOCs) in O_2 under UV–vis and visible light irradiation for mesoporous TiO_2/hydro-CF. Mesoporous TiO_2 films (brown spots) can be excited to form e^-/h^+ under UV–vis and visible light irradiation. The h^+ is one of the strong oxidant for gaseous VOCs mineralization. Moreover, TiO_2/hydro-CF substrate adsorb the VOCs molecules increasing the concentration of VOCs at the interfacial between of photocatalyst and substrate, as well as facilitate the mass transfer of VOCs due to the macroporous structure. Reproduced with permission from [82]. Copyright © 4472401418350, 2018 Elsevier.

4. Conclusions

Here an overview of recently reported synthesis protocols suited for the large- scale preparation of mesoporous TiO_2 has been provided. The reviewed syntheses have been selected to comply with the criteria of scalability, morphological and structural control, and use of safe and affordable TiO_2 precursors. Mesoporous TiO_2 as a photocatalyst offers multifold advantages such as high specific surface area, the high density of surface hydroxyl moieties, and improved absorption of UV. Several examples of TiO_2 mesoporous TiO_2 have been accounted for and discussed. Overall mesoporous TiO_2 has been demonstrated to show a great potential in the field of TiO_2-based environmental photocatalysis, also considering that currently available synthetic routes are suitable for up-scaling production of TiO_2 NPs, while still ensuring a high control over size, shape, structure and textural properties. However, great efforts are still required in order to fully elucidate the interaction with UV light, accomplish useful surface functionalization and thoroughly assess the safety issues related to the real application of nanosized TiO_2-based photocatalyst.

Author Contributions: R.C., F.P., M.L.C. conceived and drafted the work. A.T., F.P. and M.D.'E.; designed the article, acquired, analyzed, and interpreted the reports present in literature. A.A. critically revised the manuscript.

Funding: This work was partially supported by the Apulia Region Funded Project FONTANAPULIA (WOBV6K5) Italy and the European H2020 funded Project InnovaConcrete (G.A. n. 760858), PON Energy for TARANTO (ARS01_00637). PON Ricerca e Innovazione 2014–2020 (DOT1302393).

References

1. Ebele, A.J.; Abdallah, M.A.-E.; Harrad, S. Pharmaceuticals and personal care products (PPCPs) in the freshwater aquatic environment. *Emerg. Contam.* **2017**, *3*, 1–16. [CrossRef]
2. Li, J.; Liu, H.; Chen, J.P. Microplastics in freshwater systems: A review on occurrence, environmental effects, and methods for microplastics detection. *Water Res.* **2018**, *137*, 362–374. [CrossRef] [PubMed]

3. Balbuena, J.; Cruz-Yusta, M.; Sánchez, L. Nanomaterials to Combat NO X Pollution. *J. Nanosci. Nanotechnol.* **2015**, *15*, 6373–6385. [CrossRef] [PubMed]

4. Marques, J.A.; Costa, P.G.; Marangoni, L.F.B.; Pereira, C.M.; Abrantes, D.P.; Calderon, E.N.; Castro, C.B.; Bianchini, A. Environmental health in southwestern Atlantic coral reefs: Geochemical, water quality and ecological indicators. *Sci. Total Environ.* **2018**, *651*, 261–270. [CrossRef] [PubMed]

5. Fiorentino, A.; Ferro, G.; Alferez, M.C.; Polo-López, M.I.; Fernández-Ibañez, P.; Rizzo, L. Inactivation and regrowth of multidrug resistant bacteria in urban wastewater after disinfection by solar-driven and chlorination processes. *J. Photochem. Photobiol. B* **2015**, *148*, 43–50. [CrossRef] [PubMed]

6. Ângelo, J.; Andrade, L.; Madeira, L.M.; Mendes, A. An overview of photocatalysis phenomena applied to NOx abatement. *J. Environ. Manag.* **2013**, *129*, 522–539. [CrossRef] [PubMed]

7. Chong, M.N.; Jin, B.; Chow, C.W.K.; Saint, C. Recent developments in photocatalytic water treatment technology: A review. *Water Res.* **2010**, *44*, 2997–3027. [CrossRef] [PubMed]

8. Petronella, F.; Diomede, S.; Fanizza, E.; Mascolo, G.; Sibillano, T.; Agostiano, A.; Curri, M.L.; Comparelli, R. Photodegradation of nalidixic acid assisted by TiO_2 nanorods/Ag nanoparticles based catalyst. *Chemosphere* **2013**, *91*, 941–947. [CrossRef]

9. Hoffmann, M.R.; Martin, S.T.; Choi, W.; Bahnemann, D.W. Environmental Applications of Semiconductor Photocatalysis. *Chem. Rev.* **1995**, *95*, 69–96. [CrossRef]

10. Schneider, J.; Matsuoka, M.; Takeuchi, M.; Zhang, J.; Horiuchi, Y.; Anpo, M.; Bahnemann, D.W. Understanding TiO_2 Photocatalysis: Mechanisms and Materials. *Chem. Rev.* **2014**, *114*, 9919–9986. [CrossRef]

11. Petronella, F.; Truppi, A.; Sibillano, T.; Giannini, C.; Striccoli, M.; Comparelli, R.; Curri, M.L. Multifunctional TiO_2/FexOy/Ag based nanocrystalline heterostructures for photocatalytic degradation of a recalcitrant pollutant. *Catal. Today* **2017**, *284*, 100–106. [CrossRef]

12. Carp, O.; Huisman, C.L.; Reller, A. Photoinduced reactivity of titanium dioxide. *Prog. Solid State Chem.* **2004**, *32*, 33–177. [CrossRef]

13. Petronella, F.; Pagliarulo, A.; Striccoli, M.; Calia, A.; Lettieri, M.; Colangiuli, D.; Curri, M.L.; Comparelli, R. Colloidal Nanocrystalline Semiconductor Materials as Photocatalysts for Environmental Protection of Architectural Stone. *Crystals* **2017**, *7*, 30. [CrossRef]

14. Zhu, X.; Zhou, J.; Cai, Z. TiO_2 Nanoparticles in the Marine Environment: Impact on the Toxicity of Tributyltin to Abalone (Haliotis diversicolor supertexta) Embryos. *Environ. Sci. Technol.* **2011**, *45*, 3753–3758. [CrossRef] [PubMed]

15. Truppi, A.; Petronella, F.; Placido, T.; Striccoli, M.; Agostiano, A.; Curri, M.; Comparelli, R. Visible-Light-Active TiO_2-Based Hybrid Nanocatalysts for Environmental Applications. *Catalysts* **2017**, *7*, 100. [CrossRef]

16. Liu, S.; Yu, J.; Jaroniec, M. Anatase TiO_2 with Dominant High-Energy {001} Facets: Synthesis, Properties, and Applications. *Chem. Mater.* **2011**, *23*, 4085–4093. [CrossRef]

17. Ali, I.; Suhail, M.; Alothman, Z.A.; Alwarthan, A. Recent advances in syntheses, properties and applications of TiO_2 nanostructures. *RSC Adv.* **2018**, *8*, 30125–30147. [CrossRef]

18. Pang, Y.L.; Lim, S.; Ong, H.C.; Chong, W.T. A critical review on the recent progress of synthesizing techniques and fabrication of TiO_2-based nanotubes photocatalysts. *Appl. Catal. A* **2014**, *481*, 127–142. [CrossRef]

19. Reddy, K.R.; Hassan, M.; Gomes, V.G. Hybrid nanostructures based on titanium dioxide for enhanced photocatalysis. *Appl. Catal. A* **2015**, *489*, 1–16. [CrossRef]

20. Truppi, A.; Petronella, F.; Placido, T.; Margiotta, V.; Lasorella, G.; Giotta, L.; Giannini, C.; Sibillano, T.; Murgolo, S.; Mascolo, G.; et al. Gram-scale synthesis of UV—Vis light active plasmonic photocatalytic nanocomposite based on TiO_2/Au nanorods for degradation of pollutants in water. *Appl. Catal. B* **2019**, *243*, 604–613. [CrossRef]

21. Dudem, B.; Bharat, L.K.; Leem, J.W.; Kim, D.H.; Yu, J.S. Hierarchical Ag/TiO_2/Si Forest-Like Nano/Micro-Architectures as Antireflective, Plasmonic Photocatalytic, and Self-Cleaning Coatings. *ACS Sustain. Chem. Eng.* **2018**, *6*, 1580–1591. [CrossRef]

22. Veziroglu, S.; Ghori, M.Z.; Kamp, M.; Kienle, L.; Rubahn, H.G.; Strunskus, T.; Fiutowski, J.; Adam, J.; Faupel, F.; Aktas, O.C. Photocatalytic Growth of Hierarchical Au Needle Clusters on Highly Active TiO_2 Thin Film. *Adv. Mater. Interfaces* **2018**, *5*, 1800465. [CrossRef]

23. Niu, B.; Wang, X.; Wu, K.; He, X.; Zhang, R. Mesoporous Titanium Dioxide: Synthesis and Applications in Photocatalysis, Energy and Biology. *Materials* **2018**, *11*, 1910. [CrossRef] [PubMed]

24. Chen, J.; Qiu, F.; Xu, W.; Cao, S.; Zhu, H. Recent progress in enhancing photocatalytic efficiency of TiO$_2$-based materials. *Appl. Catal. A* **2015**, *495*, 131–140. [CrossRef]

25. Zhang, R.; Elzatahry, A.A.; Al-Deyab, S.S.; Zhao, D. Mesoporous titania: From synthesis to application. *Nano Today* **2012**, *7*, 344–366. [CrossRef]

26. Yu, Z.; Gao, X.; Yao, Y.; Zhang, X.; Bian, G.-Q.; Wu, W.D.; Chen, X.D.; Li, W.; Selomulya, C.; Wu, Z.; et al. Scalable synthesis of wrinkled mesoporous titania microspheres with uniform large micron sizes for efficient removal of Cr(vi). *J. Mater. Chem. A* **2018**, *6*, 3954–3966. [CrossRef]

27. Li, H.; Bian, Z.; Zhu, J.; Zhang, D.; Li, G.; Huo, Y.; Li, H.; Lu, Y. Mesoporous Titania Spheres with Tunable Chamber Stucture and Enhanced Photocatalytic Activity. *J. Am. Chem. Soc.* **2007**, *129*, 8406–8407. [CrossRef] [PubMed]

28. Kondo, Y.; Yoshikawa, H.; Awaga, K.; Murayama, M.; Mori, T.; Sunada, K.; Bandow, S.; Iijima, S. Preparation, Photocatalytic Activities, and Dye-Sensitized Solar-Cell Performance of Submicron-Scale TiO$_2$ Hollow Spheres. *Langmuir* **2008**, *24*, 547–550. [CrossRef] [PubMed]

29. Nakata, K.; Ochiai, T.; Murakami, T.; Fujishima, A. Photoenergy conversion with TiO$_2$ photocatalysis: New materials and recent applications. *Electrochim. Acta* **2012**, *84*, 103–111. [CrossRef]

30. Available online: https://statnano.com/ (accessed on 6 June 2019).

31. Yang, H.; Coombs, N.; Sokolov, I.; Ozin, G.A. Free-standing and oriented mesoporous silica films grown at the air-water interface. *Nature* **1996**, *381*, 589–592. [CrossRef]

32. Lu, Y.; Fan, H.; Stump, A.; Ward, T.L.; Rieker, T.; Brinker, C.J. Aerosol-assisted self-assembly of mesostructured spherical nanoparticles. *Nature* **1999**, *398*, 223. [CrossRef]

33. Liu, T.; Li, B.; Hao, Y.; Han, F.; Zhang, L.; Hu, L. A general method to diverse silver/mesoporous–metal–oxide nanocomposites with plasmon-enhanced photocatalytic activity. *Appl. Catal. B* **2015**, *165*, 378–388. [CrossRef]

34. Antoniou, M.G.; Nicolaou, P.A.; Shoemaker, J.A.; de la Cruz, A.A.; Dionysiou, D.D. Impact of the morphological properties of thin TiO$_2$ photocatalytic films on the detoxification of water contaminated with the cyanotoxin, microcystin-LR. *Appl. Catal. B* **2009**, *91*, 165–173. [CrossRef]

35. Liu, Y.; Wang, X.; Yang, F.; Yang, X. Excellent antimicrobial properties of mesoporous anatase TiO$_2$ and Ag/TiO$_2$ composite films. *Microporous Mesoporous Mater.* **2008**, *114*, 431–439. [CrossRef]

36. Hernández-Gordillo, A.; Camperoa, A.; Vera-Roblesb, L.I. Mesoporous TiO$_2$ synthesis using a semi-hard biological template. *Microporous Mesoporous Mater.* **2018**, *270*, 140–148. [CrossRef]

37. Yin, Q.; Xiang, J.; Wang, X.; Guo, X.; Zhang, T. Preparation of highly crystalline mesoporous TiO$_2$ by sol–gel method combined with two-step calcining process. *J. Exp. Nanosci.* **2016**, *11*, 1127–1137. [CrossRef]

38. Phattepur, H.; Siddaiah, G.B.; Ganganagappa, N. Synthesis and Characterisation of Mesoporous TiO$_2$Nanoparticles by Novel Surfactant Assisted Sol-gel Method for the Degradation of Organic Compounds. *Period. Polytech. Chem. Eng.* **2018**, *63*, 85–95. [CrossRef]

39. Ohno, T. Rutile Titanium Dioxide Nanoparticles Each Having Novel Exposed Crystal Face and Method for Producing Same. U.S. Patent Documents US 8,758,574, 24 June 2014.

40. Park; Yang, S.H.; Jun, Y.-S.; Hong, W.H.; Kang, J.K. Facile Route to Synthesize Large-Mesoporous γ-Alumina by Room Temperature Ionic Liquids. *Chem. Mater.* **2007**, *19*, 535–542. [CrossRef]

41. Yu, J.; Li, Q.; Liu, S.; Jaroniec, M. Ionic-Liquid-Assisted Synthesis of Uniform Fluorinated B/C-Codoped TiO$_2$ Nanocrystals and Their Enhanced Visible-Light Photocatalytic Activity. *Chem. Eur. J.* **2013**, *19*, 2433–2441. [CrossRef]

42. Zheng, W.; Liu, X.; Yan, Z.; Zhu, L. Ionic Liquid-Assisted Synthesis of Large-Scale TiO$_2$ Nanoparticles with Controllable Phase by Hydrolysis of TiCl4. *ACS Nano* **2009**, *3*, 115–122. [CrossRef]

43. Zhou, Y.; Antonietti, M. Synthesis of Very Small TiO$_2$ Nanocrystals in a Room-Temperature Ionic Liquid and Their Self-Assembly toward Mesoporous Spherical Aggregates. *J. Am. Chem. Soc.* **2003**, *125*, 14960–14961. [CrossRef] [PubMed]

44. Li, F.-T.; Wang, X.-J.; Zhao, Y.; Liu, J.-X.; Hao, Y.-J.; Liu, R.-H.; Zhao, D.-S. Ionic-liquid-assisted synthesis of high-visible-light-activated N–B–F-tri-doped mesoporous TiO$_2$ via a microwave route. *Appl. Catal. B* **2014**, *144*, 442–453. [CrossRef]

45. Huddleston, J.G.; Visser, A.E.; Reichert, W.M.; Willauer, H.D.; Broker, G.A.; Rogers, R.D. Characterization and comparison of hydrophilic and hydrophobic room temperature ionic liquids incorporating the imidazolium cation. *Green Chem.* **2001**, *3*, 156–164. [CrossRef]

46. Hsiung, T.-L.; Wang, H.P.; Wei, Y.-L. Preparation of Nitrogen-Doped Mesoporous TiO_2 with a Room-Temperature Ionic Liquid. In Proceedings of the Nanotech Conference Expo 2010, Anaheim, CA, USA, 21–24 June 2010; pp. 440–443.

47. Lee, H.Y.; Kale, G.M. Hydrothermal Synthesis and Characterization of Nano-TiO_2. *Int. J. Appl. Ceram. Technol.* **2008**, *5*, 657–665. [CrossRef]

48. Chen, X.; Mao, S.S. Titanium Dioxide Nanomaterials: Synthesis, Properties, Modifications, and Applications. *Chem. Rev.* **2007**, *107*, 2891–2959. [CrossRef]

49. Lee, K.-H.; Song, S.-W. One-Step Hydrothermal Synthesis of Mesoporous Anatase TiO_2 Microsphere and Interfacial Control for Enhanced Lithium Storage Performance. *ACS Appl. Mater. Interfaces* **2011**, *3*, 3697–3703. [CrossRef] [PubMed]

50. Deng, A.; Zhu, Y.; Guo, X.; Zhou, L.; Jiang, Q. Synthesis of Various TiO_2 Micro-/Nano-Structures and Their Photocatalytic Performance. *Materials* **2018**, *11*, 995. [CrossRef] [PubMed]

51. Arpaç, E.; Sayılkan, F.; Asiltürk, M.; Tatar, P.; Kiraz, N.; Sayılkan, H. Photocatalytic performance of Sn-doped and undoped TiO_2 nanostructured thin films under UV and vis-lights. *J. Hazard. Mater.* **2007**, *140*, 69–74. [CrossRef]

52. Querejeta, A.; Varela, A.; Parras, M.; del Monte, F.; García-Hernández, M.; González-Calbet, J.M. Hydrothermal Synthesis: A Suitable Route to Elaborate Nanomanganites. *Chem. Mater.* **2009**, *21*, 1898–1905. [CrossRef]

53. Anwar, M.S.; Danish, R.; Ahmed, F.; Koo, B.H. Pressure Dependent Synthesis and Enhanced Photocatalytic Activity of TiO_2 Nano-Structures. *Nanosci. Nanotechnol. Lett.* **2016**, *8*, 778–781. [CrossRef]

54. Kolen'ko, Y.V.; Maximov, V.D.; Garshev, A.V.; Meskin, P.E.; Oleynikov, N.N.; Churagulov, B.R. Hydrothermal synthesis of nanocrystalline and mesoporous titania from aqueous complex titanyl oxalate acid solutions. *Chem. Phys. Lett.* **2004**, *388*, 411–415. [CrossRef]

55. Galarneau, A.; Cambon, H.; Di Renzo, F.; Ryoo, R.; Choi, M.; Fajula, F. Microporosity and connections between pores in SBA-15 mesostructured silicas as a function of the temperature of synthesis. *New J. Chem.* **2003**, *27*, 73–79. [CrossRef]

56. Wang, Y.; Jiang, Z.-H.; Yang, F.-J. Preparation and photocatalytic activity of mesoporous TiO_2 derived from hydrolysis condensation with TX-100 as template. *Mater. Sci. Eng. B* **2006**, *128*, 229–233. [CrossRef]

57. Kim, D.S.; Kwak, S.-Y. The hydrothermal synthesis of mesoporous TiO_2 with high crystallinity, thermal stability, large surface area, and enhanced photocatalytic activity. *Appl. Catal. A* **2007**, *323*, 110–118. [CrossRef]

58. Zhou, M.; Xu, J.; Yu, H.; Liu, S. Low-temperature hydrothermal synthesis of highly photoactive mesoporous spherical TiO_2 nanocrystalline. *J. Phys. Chem. Solids* **2010**, *71*, 507–510. [CrossRef]

59. Santhosh, N.; Govindaraj, R.; Senthil Pandian, M.; Ramasamy, P.; Mukhopadhyay, S. Mesoporous TiO_2 microspheres synthesized via a facile hydrothermal method for dye sensitized solar cell applications. *J. Porous Mater.* **2016**, *23*, 1483–1487. [CrossRef]

60. Cui, Y.; Liu, L.; Li, B.; Zhou, X.; Xu, N. Fabrication of Tunable Core–Shell Structured TiO_2 Mesoporous Microspheres Using Linear Polymer Polyethylene Glycol as Templates. *J. Phys. Chem. C* **2010**, *114*, 2434–2439. [CrossRef]

61. Ye, M.; Chen, Z.; Wang, W.; Shen, J.; Ma, J. Hydrothermal synthesis of TiO_2 hollow microspheres for the photocatalytic degradation of 4-chloronitrobenzene. *J. Hazard. Mater.* **2010**, *184*, 612–619. [CrossRef] [PubMed]

62. Liu, Z.; Sun, D.D.; Guo, P.; Leckie, J.O. One-Step Fabrication and High Photocatalytic Activity of Porous TiO_2 Hollow Aggregates by Using a Low-Temperature Hydrothermal Method Without Templates. *Chem. Eur. J.* **2007**, *13*, 1851–1855. [CrossRef] [PubMed]

63. Sattarfard, R.; Behnajady, M.A.; Eskandarloo, H. Hydrothermal synthesis of mesoporous TiO_2 nanotubes and their adsorption affinity toward Basic Violet 2. *J. Porous Mater.* **2018**, *25*, 359–371. [CrossRef]

64. Liu, N.; Chen, X.; Zhang, J.; Schwank, J.W. A review on TiO_2-based nanotubes synthesized via hydrothermal method: Formation mechanism, structure modification, and photocatalytic applications. *Catal. Today* **2014**, *225*, 34–51. [CrossRef]

65. Asiah, M.N.; Mamat, M.H.; Khusaimi, Z.; Abdullah, S.; Rusop, M.; Qurashi, A. Surfactant-free seed-mediated large-scale synthesis of mesoporous TiO_2 nanowires. *Ceram. Int.* **2015**, *41*, 4260–4266. [CrossRef]

66. Guo, C.; Ge, M.; Liu, L.; Gao, G.; Feng, Y.; Wang, Y. Directed Synthesis of Mesoporous TiO_2 Microspheres: Catalysts and Their Photocatalysis for Bisphenol A Degradation. *Environ. Sci. Technol.* **2010**, *44*, 419–425. [CrossRef] [PubMed]

67. Pan, J.H.; Wang, X.Z.; Huang, Q.; Shen, C.; Koh, Z.Y.; Wang, Q.; Engel, A.; Bahnemann, D.W. Large-scale Synthesis of Urchin-like Mesoporous TiO_2 Hollow Spheres by Targeted Etching and Their Photoelectrochemical Properties. *Adv. Funct. Mater.* **2014**, *24*, 95–104. [CrossRef]

68. Baloyi, J.; Seadira, T.; Raphulu, M.; Ochieng, A. Preparation, Characterization and Growth Mechanism of Dandelion-like TiO_2 Nanostructures and their Application in Photocatalysis towards Reduction of Cr(VI). *Mater. Today: Proc.* **2015**, *2*, 3973–3987. [CrossRef]

69. Hu, C.; Lei, E.; Zhao, D.; Hu, K.; Cui, J.; Xiong, Q.; Liu, Z. Controllable synthesis and formation mechanism of 3D flower-like TiO_2 microspheres. *J. Mater. Sci.: Mater. Electron.* **2018**, *29*, 10277–10283. [CrossRef]

70. Asuha, S.; Zhou, X.G.; Zhao, S. Adsorption of methyl orange and Cr(VI) on mesoporous TiO_2 prepared by hydrothermal method. *J. Hazard. Mater.* **2010**, *181*, 204–210. [CrossRef] [PubMed]

71. Zhang, F.; Zhang, Y.; Song, S.; Zhang, H. Superior electrode performance of mesoporous hollow TiO_2 microspheres through efficient hierarchical nanostructures. *J. Power Sources* **2011**, *196*, 8618–8624. [CrossRef]

72. Dionysiou, D.D.; Khodadoust, A.P.; Kern, A.M.; Suidan, M.T.; Baudin, I.; Laîné, J.-M. Continuous-mode photocatalytic degradation of chlorinated phenols and pesticides in water using a bench-scale TiO_2 rotating disk reactor. *Appl. Catal. B* **2000**, *24*, 139–155. [CrossRef]

73. Ozyonar, F.; Aksoy, S.M. Removal of Salicylic Acid from Aqueous Solutions Using Various Electrodes and Different Connection Modes by Electrocoagulation. *Int. J. Electrochem. Sci.* **2016**, *11*, 3680–3696. [CrossRef]

74. Halling-Sørensen, B.; Nors Nielsen, S.; Lanzky, P.F.; Ingerslev, F.; Holten Lützhøft, H.C.; Jørgensen, S.E. Occurrence, fate and effects of pharmaceutical substances in the environment—A review. *Chemosphere* **1998**, *36*, 357–393. [CrossRef]

75. Chowdhury, I.H.; Naskar, M.K. Sol-gel synthesis of mesoporous hollow titania microspheres for photodegradation of 4-chlorophenol. *Indian J. Chem.* **2018**, *57A*, 910–914.

76. Petronella, F.; Truppi, A.; Ingrosso, C.; Placido, T.; Striccoli, M.; Curri, M.L.; Agostiano, A.; Comparelli, R. Nanocomposite materials for photocatalytic degradation of pollutants. *Catal. Today* **2017**, *281*, 85–100. [CrossRef]

77. Shayegan, Z.; Lee, C.-S.; Haghighat, F. TiO_2 photocatalyst for removal of volatile organic compounds in gas phase—A review. *Chem. Eng. J.* **2018**, *334*, 2408–2439. [CrossRef]

78. Mendoza, C.; Valle, A.; Castellote, M.; Bahamonde, A.; Faraldos, M. TiO_2 and TiO_2–SiO2 coated cement: Comparison of mechanic and photocatalytic properties. *Appl. Catal. B* **2015**, *178*, 155–164. [CrossRef]

79. Kalousek, V.; Tschirch, J.; Bahnemann, D.; Rathouský, J. Mesoporous layers of TiO_2 as highly efficient photocatalysts for the purification of air. *Superlattices Microstruct.* **2008**, *44*, 506–513. [CrossRef]

80. Rathouský, J.; Kalousek, V.; Yarovyi, V.; Wark, M.; Jirkovský, J. A low-cost procedure for the preparation of mesoporous layers of TiO_2 efficient in the environmental clean-up. *J. Photochem. Photobiol. A-Chem.* **2010**, *216*, 126–132. [CrossRef]

81. Balbuena, J.; Calatayud, J.M.; Cruz-Yusta, M.; Pardo, P.; Martín, F.; Alarcón, J.; Sánchez, L. Mesocrystalline anatase nanoparticles synthesized using a simple hydrothermal approach with enhanced light harvesting for gas-phase reaction. *Dalton Trans.* **2018**, *47*, 6590–6597. [CrossRef]

82. Qian, X.; Ren, M.; Yue, D.; Zhu, Y.; Han, Y.; Bian, Z.; Zhao, Y. Mesoporous TiO_2 films coated on carbon foam based on waste polyurethane for enhanced photocatalytic oxidation of VOCs. *Appl. Catal. B* **2017**, *212*, 1–6. [CrossRef]

83. Ji, J.; Xu, Y.; Huang, H.; He, M.; Liu, S.; Liu, G.; Xie, R.; Feng, Q.; Shu, Y.; Zhan, Y.; et al. Mesoporous TiO_2 under VUV irradiation: Enhanced photocatalytic oxidation for VOCs degradation at room temperature. *Chem. Eng. J.* **2017**, *327*, 490–499. [CrossRef]

84. Ochiai, T.; Fujishima, A. Photoelectrochemical properties of TiO_2 photocatalyst and its applications for environmental purification. *J. Photochem. Photobiol. C-Photochem. Rev.* **2012**, *13*, 247–262. [CrossRef]

Article

Photocatalytic Degradation of Diclofenac by Hydroxyapatite–TiO$_2$ Composite Material: Identification of Transformation Products and Assessment of Toxicity

Sapia Murgolo [1,†], Irina S. Moreira [2,†], Clara Piccirillo [3], Paula M. L. Castro [2], Gianrocco Ventrella [1,4], Claudio Cocozza [4] and Giuseppe Mascolo [1,*]

1   CNR, Istituto di Ricerca Sulle Acque, Via F. De Blasio 5, 70132 Bari, Italy; sapia.murgolo@ba.irsa.cnr.it (S.M.); gianrocco.ventrella@uniba.it (G.V.)
2   CBQF—Centro de Biotecnologia e Química Fina, Laboratório Associado, Escola Superior de Biotecnologia, Universidade Católica Portuguesa/Porto, Rua Arquiteto Lobão Vital, 172, 4200-374 Porto, Portugal; ismoreira@porto.ucp.pt (I.S.M.); plcastro@porto.ucp.pt (P.M.L.C.)
3   CNR, Institute of Nanotechnology, Campus Ecoteckne, Via Monteroni, 73100 Lecce, Italy; clara.piccirillo@nanotec.cnr.it
4   Dipartimento di Scienze del Suolo, della Pianta e degli Alimenti—Di.S.S.P.A., Università di Bari, Via Amendola 165/A, 70126 Bari, Italy; claudio.cocozza@uniba.it
*   Correspondence: giuseppe.mascolo@ba.irsa.cnr.it; Tel.: +39-080-5820519
†   These authors equally contributed to this work.

Received: 24 August 2018; Accepted: 17 September 2018; Published: 19 September 2018

Abstract: Diclofenac (DCF) is one of the most detected pharmaceuticals in environmental water matrices and is known to be recalcitrant to conventional wastewater treatment plants. In this study, degradation of DCF was performed in water by photolysis and photocatalysis using a new synthetized photocatalyst based on hydroxyapatite and TiO$_2$ (HApTi). A degradation of 95% of the target compound was achieved in 24 h by a photocatalytic treatment employing the HApTi catalyst in comparison to only 60% removal by the photolytic process. The investigation of photo-transformation products was performed by means of UPLC-QTOF/MS/MS, and for 14 detected compounds in samples collected during treatment with HApTi, the chemical structure was proposed. The determination of transformation product (TP) toxicity was performed by using different assays: *Daphnia magna* acute toxicity test, Toxi-ChromoTest, and *Lactuca sativa* and *Solanum lycopersicum* germination inhibition test. Overall, the toxicity of the samples obtained from the photocatalytic experiment with HApTi decreased at the end of the treatment, showing the potential applicability of the catalyst for the removal of diclofenac and the detoxification of water matrices.

Keywords: diclofenac; hydroxyapatite; photocatalysis; transformation products; toxicity

---

1. Introduction

The most common wastewater treatment plants (WWTPs) used worldwide are mainly based on the activated sludge technique and are not able to remove and/or degrade several families of trace compounds, known as "contaminants of emerging concern" (CECs). Particularly, the use of pharmaceuticals (PhACs) in everyday human life, continually introduced from the market for the healthy population and easily accessible without a prescription, represents a relevant source of contamination of the aquatic environment, as several of these are not completely metabolized by the human organism; they are, therefore, eliminated by urine and faeces [1–3].

Moreover, biological processes in WWTPs as well as subsequent chemical treatments may convert PhACs into transformation products (TPs) that can be more toxic than their parent compounds [4–6].

Several toxicological and ecotoxicological studies have been performed to obtain an overview of the risk assessment of TPs detected in waters and wastewaters [7–9].

Both trace compounds and their relative TPs have been found in different environmental compartments, including WWTP effluent, surface water, groundwater, drinking water, soil, sediments, and sludge, and their concentration ranged from μg L^{-1} to ng L^{-1}. At present, CECs are not regulated, but the relevance of addressing this issue was acknowledged by the Directive 2013/39/EU listing priority substances and further supported by the implementation of Decision (EU) 2015/495 [10]. In addition, no relevant knowledge is available about the potential ecological impacts of CECs. The continuous discharge of and long-term exposure to these contaminants, however, may have long-term effects on human health. Reproductive impairment, hormonal dysfunction, development of antibiotic-resistant bacteria, and cancer may be possible consequences [11,12].

Among the PhACs, biodegradation of nonsteroidal anti-inflammatory drugs (NSAIDs) has been found to be slow or negligible. Several reviews published in recent years have discussed the occurrence, persistence, and fate of the diclofenac (DCF). DCF is one of the most detected PhACs in the water matrix and it is recalcitrant to conventional wastewater treatment plants. Data reported in the literature have shown that WWTPs remove DCF with an efficiency in the range of 21–40% [13–16].

After consumption, DCF is mostly transformed into hydroxyl metabolites from human metabolism and subsequently excreted from the body in the aqueous environment. Furthermore, more DCF metabolites are present in aqueous matrices due to biological processes in WWTPs which convert DCF to hydroxydiclofenac; the same process takes place through a photochemical process, i.e., photolysis. These data highlight that not only DCF but also its metabolites and photo-transformation products with different physicochemical properties can be sources of contamination in the aqueous environment [17,18].

The literature reports that the combination of biological processes with advanced oxidation processes (AOPs) led to increased efficiency in the removal of organics [19,20]; because of this, a great amount of work has been published on AOPs as efficient treatments for the removal of PhACs and their metabolites in water [21–23]. Among the AOPs, particular attention has been given to heterogeneous photocatalysis based on materials such as TiO$_2$, as they represent a promising and environmentally sustainable wastewater treatment technology. Their activity is based on the formation of highly reactive OH radicals which are nonselective and have a high oxidative power (E0 = +2.80 V) [24,25]. In the field of photocatalysis, a large number of new TiO$_2$-based photocatalysts have been developed combining TiO$_2$ with other materials that can enhance its performance [26–29]. The combination with hydroxyapatite (Ca$_{10}$(PO$_4$)(OH)$_2$, HAp), in particular, seems very promising due to its excellent absorption properties. In fact, in a multiphasic HAp-TiO$_2$ material, pollutants get adsorbed on the surface more easily, leading to a higher degradation rate ([30] and references therein).

A recent study showed that it is possible to obtain a photocatalyst based on HAp and TiO$_2$ using cod fish bones as a source; HAp, in fact, is the main component in human and animal bones. Piccirillo et al. showed that treating the bones in a Ti (IV)-containing solution and successively calcining them leads to a photoactive multiphasic material [31]. This material was tested in the degradation of DCF in aqueous matrices using a bench-scale system [32]. Results showed that when compared to photolysis, HAp-titania significantly increases the diclofenac removal and is also effective when employed in a complex matrix (wastewater) or when reused in subsequent cycles of treatments. FTIR measurements of the photocatalyst taken before and after the photodegradation experiments showed that the powder material did not present any significant property change [32].

However, measurements of total organic carbon at the end of the experiments showed incomplete mineralization of the target pollutant. This indicates the production of some TPs which can be less biodegradable and/or more toxic than the parent compound.

Previous studies have investigated diclofenac TPs in water matrices after photocatalytic degradation [33–35]; to the best of our knowledge, however, no work has been done when hydroxyapatite-based catalysts were used.

The aim of this work is, therefore, a follow-up of the experimental results described in E. Màrquez Brazòn et al.'s work [32] in order to investigate the possible presence of photo-transformation products of diclofenac in water after photolysis and photocatalysis with hydroxyapatite and TiO_2 (HApTi) and to assess their level of toxicity in comparison with the parent DCF.

The investigation of TP structure was performed by means of an analytical approach based on high-resolution mass spectrometry coupled with liquid chromatography (LC-HRMS), providing a proposed molecular formula and elucidating the respective chemical structures [36]. Considering the proposed chemical structures of the identified TPs, the degradation mechanism of DCF with only photolysis and with photocatalysis (in the presence of a hydroxyapatite-based catalyst) was compared. Finally, for a complete risk valuation study on TPs of DCF formed during the investigated treatments, determination of their toxicity was performed by using different assays: *Daphnia magna* acute toxicity test, Toxi-ChromoTest, and *Lactuca sativa* and *Solanum lycopersicum* germination inhibition test.

2. Materials and Methods

2.1. Chemicals

DCF was purchased from Sigma-Aldrich (Steinheim, Germany). HPLC-grade methanol (Riedel-de Haën, Baker) was used to prepare a stock standard solution of DCF as well as for ultra-performance liquid chromatography (UPLC) analysis. Ultrapure water (18.2 MΩcm, organic carbon ≤ 4 µg/L) was supplied by a Milli-Q water system (Gradient A-10, Millipore SAS, Molsheim, France) and used for both UPLC analysis and to perform investigated treatments. All the analysed samples were previously filtered on syringe CA membrane filters with a 0.45-µm pore size.

2.2. Synthesis and Characterisation of HApTi

A detailed description of the photocatalytic material preparation and characterization was previously published [31]. Briefly, the photocatalyst was prepared from cod fish bones, which were treated in a basic $Ti(SO_4)_2$ solution and successively calcined at 800 °C. The powder was constituted by hydroxyapatite (HAp), tricalcium phosphate, and TiO_2 (anatase) in a proportion of 54, 45, and 1 wt %, respectively. For a more complete description of the material, see the reference literature.

2.3. Bench-Scale Experiments

Photocatalysis experiments were performed as previously described [32]. Briefly, a volume of 50 mL of DCF solution at 5 mg/L in distilled water was placed in 100-mL flasks with 0.2 g of photocatalyst to achieve a concentration of 4 g/L. The flasks were magnetically stirred and irradiated from the top with a XX-15 BLB UV lamp (λ 365 nm) (see reference [37] for the full emission spectrum); the irradiation density was 1.80 mW/cm^2 (value given by the producer). Experiments were also performed as described above but with no photocatalyst to monitor the DCF degradation due to UV illumination. Before starting the experiments, the solutions were left stirring in the dark for 30 min to ensure that removal of DCF was due to photodegradation and not to adsorption on the surface of the photocatalyst. Samples were taken at regular intervals to assess the degradation of DCF. DCF concentration was determined by a validated HPLC-UV method [38,39]. Experiments were performed in duplicate. New experiments were performed in triplicate for the identification of the TPs. Samples were analysed by UPLC-QTOF/MS/MS as described in the following section.

2.4. Analytical Setup and Data Processing

An Ultimate 3000 System (Thermo Fisher Scientific, Waltham, MA, USA) interfaced with a TripleTOF 5600+ high-resolution mass spectrometer (AB-Sciex), equipped with a duo-spray ion source operated in electrospray (ESI) mode (in positive and negative ion modes), was used to identify TPs. An information-dependent acquisition (IDA) method was used.

The mass spectrometer was calibrated using standards recommended by AB SCIEX for calibrating the AB SCIEX TripleTOF® 5600 Instrument, i.e., ESI Positive Calibration Solution and ESI Negative Calibration Solution. These solutions consist of a list of TOF MS Calibration Ions in the range of 140–1500 Da, including reserpine and sulfinpyrazone for MS/MS calibration in positive and negative mode, respectively, for which the list of product ions is known. All the information related to these lists are saved in reference tables available in the calibration section of the instrument. Briefly, before a run of analyses, a manual calibration was carried out by injecting the calibration solution by means of a Calibrant Delivery System (CDS) in tuning mode. Each calibration solution was configured on the CDS by a specific valve position. The instrument was therefore calibrated by comparing the exact mass of the acquired ions (experimental mass) with respect to that of the ions listed in the reference tables (theoretical mass), both in TOF MS and Product Ion MS/MS mode. When the batch was submitted, the calibration samples were inserted into the queue every five samples. Each run started with a calibration sample. With the CDS configured, the software automatically created a calibration method that matched the acquisition method that was used for the next sample in the queue. Calibration data were saved to a separate data file for each calibration sample.

The MS interface conditions for sample acquisition were the following: curtain gas (psig) 35, ion source gas 1 nebulizer gas (psig) 55, ion source gas 2 turbo gas (psig) 55, IonSpray voltage (V) 5500 (positive mode) and -4500 (negative mode), source temperature ($^\circ$C) 450, declustering potential (V) 80, m/z range 100–1000 Da.

A Waters BEH C18 column 2.1 $\times$ 150 mm, 1.7 µm, was used for the chromatographic separation, operating at a flow of 0.200 mL/min. Five-microliter samples were injected and eluted with a binary gradient consisting of 1.5 mM ammonium acetate in H_2O/MeOH 95/5 (A) and 1.5 mM ammonium acetate in methanol (B) as follows: 0% B at the initial point, linearly increased to 95% in 12 min, and held for 5 min. A 7-min equilibration step at 0% B was used at the end of each run to bring the total run time per sample to 24 min. As for the data processing, both suspect target and nontarget screenings, with identification of already known in the literature and novel TPs, respectively, were carried out. Specifically, all the collected samples corresponding to the reaction times for a specific treatment were processed by a specific analytical protocol, including two steps:

- Suspect screening: The AB-Sciex software, i.e., SciexOS 1.2, PeakView 2.2, MasterView 1.1, and LibraryView 1.1.0, were employed by using a list of likely TPs collected from the literature or from prediction models. The samples were screened for those candidates on the basis of the mass exact, isotopic pattern, fragmentation MS/MS patter, and chromatographic retention time. However, since no reference standards are available for all revealed TPs, the subsequent confirmation of the analytes is not completely possible. Therefore, the molecular formula and structure of suspected molecules can be only predicted.
- Nontarget screening: An open source software, i.e., enviMass 3.5 [40], was used for the investigation of compounds for which no previous knowledge is available and which is usually carried out after suspect screening. Briefly, after a first step of peak picking, the following steps include the removal of peaks found also in the blank sample, the mass recalibration, and the componentization of isotopes and adducts.

The assignment of the molecular formula to the accurate mass of the selected peak was performed using AB-Sciex software. Moreover, based on the interpretation of the fragmentation pattern generated in MS/MS acquisition, a candidate structure was assigned to the identified TPs, which provided a tool to hypothesize a possible degradation pathway.

2.5. Toxicity Test

The toxicity of DCF solutions was evaluated through bioassays in samples collected at the times 0, 4, 6, and 24 h in the duplicate experiments, initially performed both with and without photocatalyst for assessing the DCF decay. Samples were analysed without dilution in order to assess the evolution of whole sample toxicity during photodegradation experiments and to compare the resulted final toxicity in the presence and absence of the photocatalytic material. All toxicity assays were performed in quadruplicate of combined duplicate samples from photodegradation experiments.

The 24–48-h immobilization of crustaceans *D. magna* bioassays were performed using Daphtoxkit F Magna (MicroBio Tests, Mariakerke (Gent), Belgium) following manufacturer recommendations. The toxicity was measured as the immobilization of *D. magna* according to the procedures of OCED Guideline 202 [41] and ISO 6341 [42] after 15-min exposure.

The Toxi-ChromoTest (EBPI, Mississauga, ON, Canada) was used according to manufacturer instructions to determine the samples' potential for the inhibition of the de novo synthesis of an inducible β-galactosidase by an *E. coli* mutant. The activity of the enzyme was detected by the hydrolysis of a chromogenic substrate.

The acute bioassay with *L. sativa* and *S. lycopersicum* evaluated the potential toxicity considering the inhibition of seed germination and root and shoot elongation according to OECD Guideline 208 [43]. Experiments were performed in triplicate at 25 °C for 7 days. The inhibition normalized on negative control data were expressed as percentage of effect.

3. Results and Discussion

In the present work, the experimental setup used for photodegradation was slightly changed compared to previous experiments [32] to have a higher irradiation dose and, hence, higher degradation efficiency. The results of photodegradation of DCF plotted in Figure 1 showed that in the photolytic assay, i.e., containing no catalyst, a removal of about 60% of the target compound was observed, and the presence of the HApTi catalyst increased the degradation to 95% at the end of the experiment, demonstrating the photocatalytic efficiency of the catalyst for DCF removal.

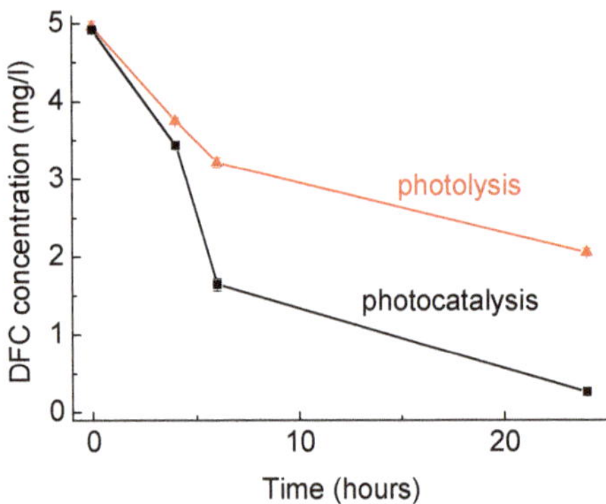

Figure 1. Diclofenac (DCF) photodegradation over time by UV photolysis (no photocatalyst) and UV photocatalysis employing hydroxyapatite and TiO$_2$ (HApTi) catalyst in solution (DCF concentration 5 ppm; HApTi catalyst concentration 4 g/L). Error bars refer to standard deviation of duplicate samples.

3.1. Identification of the Transformation Products Produced by DCF Photodegradation

Although almost 95% and 60% of DCF was removed under the investigated treatment in the presence of HApTi and only UV light, respectively. The TOC measurements showed that the pharmaceutical was not completely mineralized [32]. From this perspective, samples collected at different irradiation times (0, 4, 6, and 24 h) during both photolytic and photocatalytic experiments were analysed for TPs by UPLC/ESI-QTOF-MS-MS. The acquired chromatograms showed several peaks, of which only the main ones were identified by merging output results of both the suspect and nontarget screening protocol. In the first step of identification, the samples were processed by querying the detected peaks with a list of the main TPs collected from already published works in which DCF photodegradation was investigated [15,33,44–47]. Based on the analyses acquired in full scan mode as well as on the corresponding exact masses, considering both the isotopic cluster of molecular ions and the fragmentation pattern, 10 major photo-transformation products were identified. The analytical protocol employed for nontarget screening also allowed the detection of four new photo-transformation products clearly showing in the mass spectra the characteristic isotopic cluster of chlorine-containing compounds.

In Table 1, the accurate masses (calculated and measured) of the 14 detected Photo TPs, the ionization mode acquisition, the predicted formula, and the calculated mass errors are listed. The accurate mass results were found with an error in a range between −2 and +1 ppm, thus providing the assignment of proposed elemental compositions with confidence. For each potential photo-transformation product, the elucidation of the structure was assessed based on the interpretation of the accurate mass of the MS/MS fragments as well as considering all the reaction mechanisms likely to occur in the oxidation processes applied for DCF removal, i.e., photocatalysis using HApTi as a catalyst (Figure 2). Oxidation is one of the major degradation mechanisms during photocatalytic treatment. Indeed, the formation of oxidized compounds was confirmed by the identification of Photo TPs, the predicted formula of which showed an increase of oxygen atoms with respect to the parent DCF elemental composition. An oxidative displacement of chloride from DCF was proposed for Photo TP-1, with a decrease of 18 Da in terms of m/z resulting from the loss of HCl and subsequent hydroxylation, whereas an increase of 16 Da was indicative of DCF hydroxylation in Photo TP-2. Although MS data were not enough to define the exact position of the hydroxyl group, the fragment m/z 166.0643 detected on Photo TP-2 (Table 1) proved that the hydroxylation occurred on the nonchlorinated ring.

Table 1. DCF photo-transformation products detected by UPLC/ESI–QTOF–MS IDA.

Compounds	Ionization Mode	Calculated m/z	Measured m/z	ppm Error	Products MS/MS	Predicted Formula	Ref.
Photo TP-1	ESI (+)	278.0579	278.0577	−0.6	168.0794, 196.0755, 232.0508, 260.0470	C14H12ClNO3	[44]
Photo TP-2	ESI (+)	312.0189	312.0186	−1.0	166.0643, 194.0612, 230.0357, 265.9974	C14H11Cl2NO3	[44,47]
Photo TP-3	ESI (+)	310.0032	310.0031	−0.4	166.0657, 201.0345, 263.9987, 291.9941	C14H9Cl2NO3	[44,47]
Photo TP-4	ESI (−)	323.9836	323.9835	−0.3	152.0507, 208.0423, 252.0300, 280.0045	C14H9Cl2NO4	[15]
Photo TP-5	ESI (+)	282.0083	282.0084	0.4	166.0646, 194.0598, 229.0285, 263.9979	C13H9Cl2NO2	-
Photo TP-6	ESI (−)	266.0145	266.0147	0.8	127.0543, 166.0662, 184.0961, 206.0185	C13H11Cl2NO	[44,45]
Photo TP-7	ESI (+)	276.0422	276.0420	−0.8	166.0650, 194.0597, 202.0424, 230.0360	C14H10ClNO3	-
Photo TP-8	ESI (−)	246.0327	246.0329	0.7	141.0214, 164.0516, 200.0265, 228.0234	C13H10ClNO2	-
Photo TP-9	ESI (−)	230.0378	230.0378	−0.1	143.113, 166.0646, 194.0606, 215.0134	C13H10ClNO	-
Photo TP-10	ESI (+)	260.0473	260.0468	−1.8	125.0442, 151.0545, 165.0902, 179.0732	C14H10ClNO2	[33,45]
Photo TP-11	ESI (−)	214.0430	214.0431	1.1	65.9613, 138.0405, 142.9975, 178.0652	C13H10ClN	[45]
Photo TP-12	ESI (−)	240.0666	240.0667	0.3	99.9485, 142.0667, 168.0815, 196.0768	C14H11NO3	[33,45,46]
Photo TP-13	ESI (+)	256.0604	256.0604	−0.1	95.0885, 127.0543, 182.0595, 210.0552	C14H9NO4	[15]
Photo TP-14	ESI (−)	196.0768	196.0768	0.2	59.0159, 135.0128, 152.0321, 168.0804	C13H11NO	[45]

Photo TP-3 revealed two fewer hydrogen atoms with respect to Photo TP-2 with the formation of an additional double bound. The presence of the carboxylic acid functionality confirmed by the MS/MS fragment (fragment showed the loss of CO_2) led to assigning the structure of a benzoquinone imine species to the Photo TP-3. For Photo TP-4, a mass increase of 16 Da in relation to Photo TP-3 reveals a subsequent addition of one oxygen atom; according with the fragmentation pattern, this could be explain by a hydroxylation on the chlorinated ring. The Photo TP-5 (MW 281 Da) and Photo TP-6 (MW 267 Da) were rationalized to be formed from Photo TP-4 and Photo TP-2, respectively, both by a decarboxylation reaction of the aliphatic chain.

From plotting the time profiles of the identified Photo TPs (Figure 3), it is evident that Photo TP-1, TP-2, TP-3, TP-4, TP-5, and TP-6 are formed only during the photocatalytic treatment in the presence of HApTi, with an intensity that significantly increased during the first hours of treatment (Figure 3a–f).

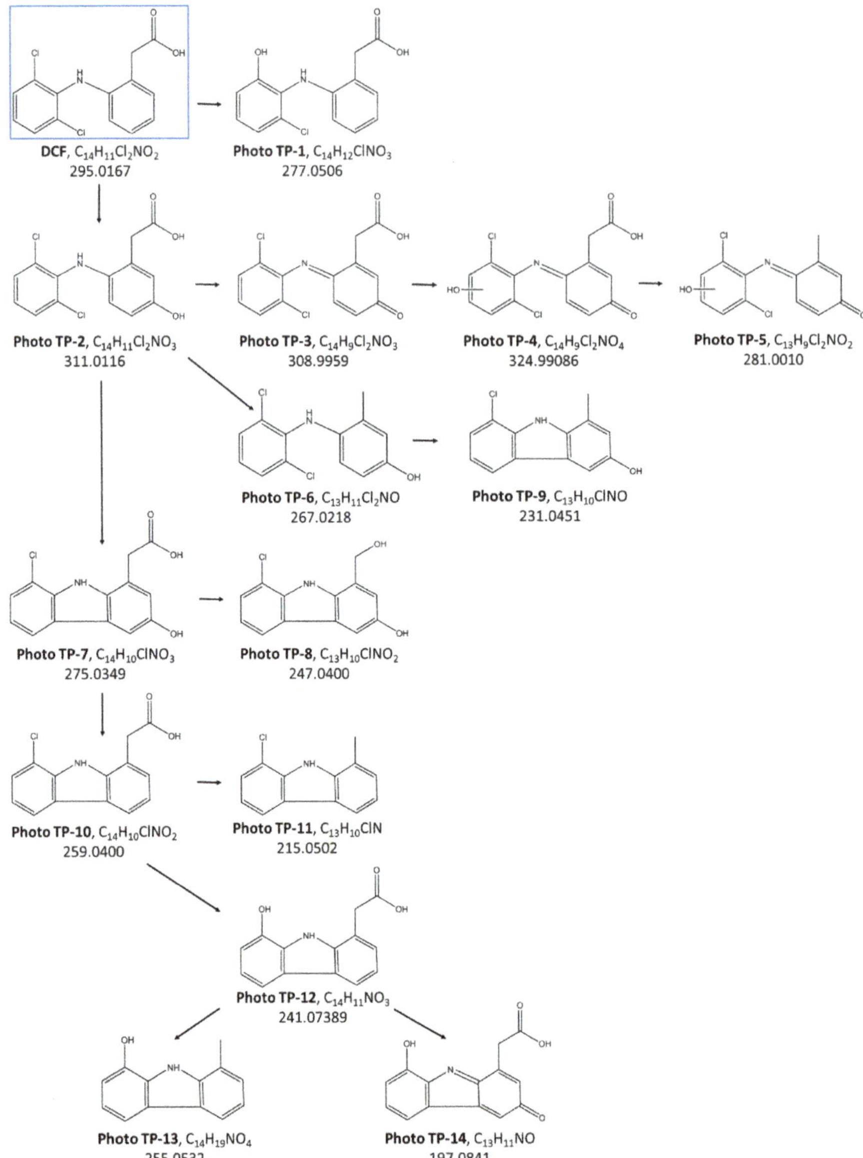

Figure 2. DCF degradation pathway proposed for the Photo TPs identified during treatment with HApTi.

The loss of one chlorine atom from Photo TP-1 and the subsequent cyclization reaction, forming a five-membered ring with nitrogen, generated Photo TP-7. The time profiles show that TP-7 formation increases with the time of photocatalytic treatment, as does Photo TP-9, which forms from Photo TP-6 by a loss of one chlorine atom (Figure 3g,i). Photo TP-8 is formed both during photolysis and photocatalysis, but only the latter treatment was able to quickly remove it (Figure 3h).

As for Photo TP-10, the isotopic pattern of which indicates the presence of only one chlorine atom, a chlorocarbazole acetic acid structure was proposed. Considering the mass decrease of 44 Da between Photo TP-10 and Photo TP-11 and the predicted formula, Photo TP-11 corresponds to the loss of the carboxylic acid group. Photo TP-12 revealed a mass decrease of 18 Da with respect to Photo TP10; the absence of the typical isotopic pattern of chlorine and the assigned formula for the detected compound could be attributed to the replacement of the chloride atom with the OH group.

The time profiles of these products (Figure 3j–l) reveal no significant difference between photolysis and photocatalysis. In both treatments, the detected TPs achieve the maximum of formation after 4 h of treatment with a subsequent decay. Although the removal is not complete at the end of the treatments, it was slightly faster in the presence of HApTi.

Two more products of MW 255 and 197 Da were observed and the molecular ion isotopic cluster indicated that no chlorine atom was present in both of these products, Photo TP-13, and Photo TP-14, respectively. The photolytic treatment generates Photo TP-13 with higher intensity with respect to photocatalytic treatment, with a subsequently slower removal of the compound over time (Figure 3m), whereas a complete removal for Photo TP-14 at the end of both treatments was observed.

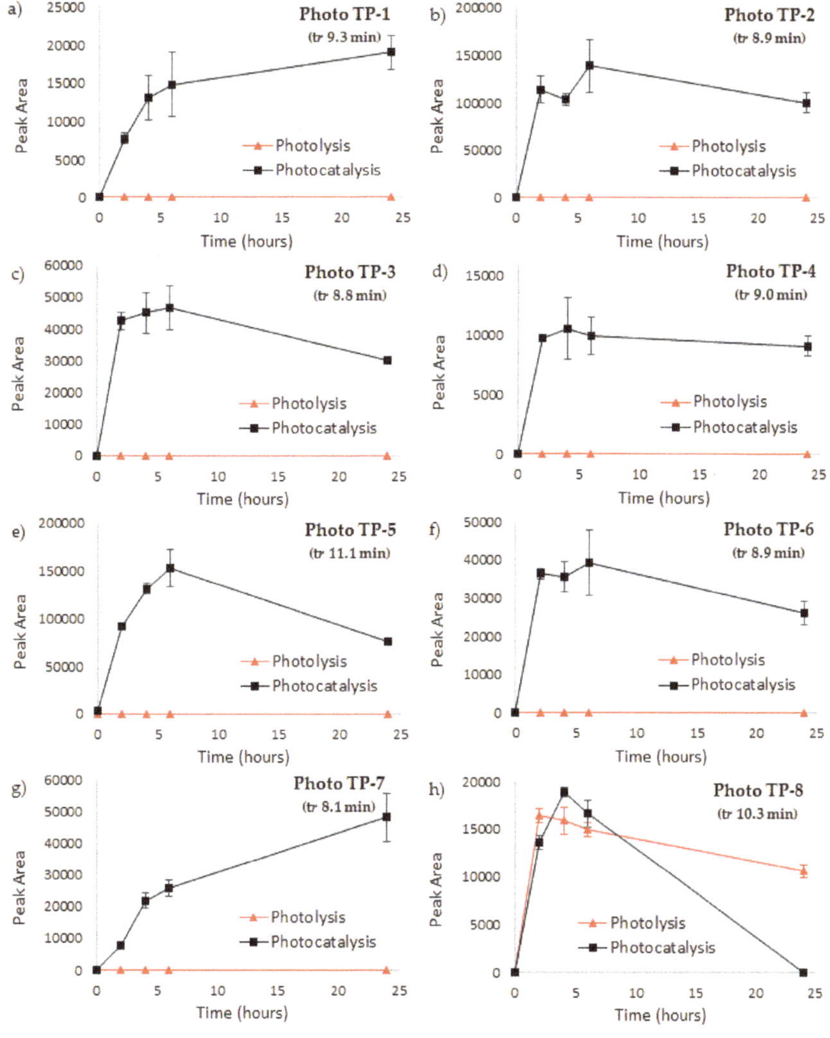

Figure 3. *Cont.*

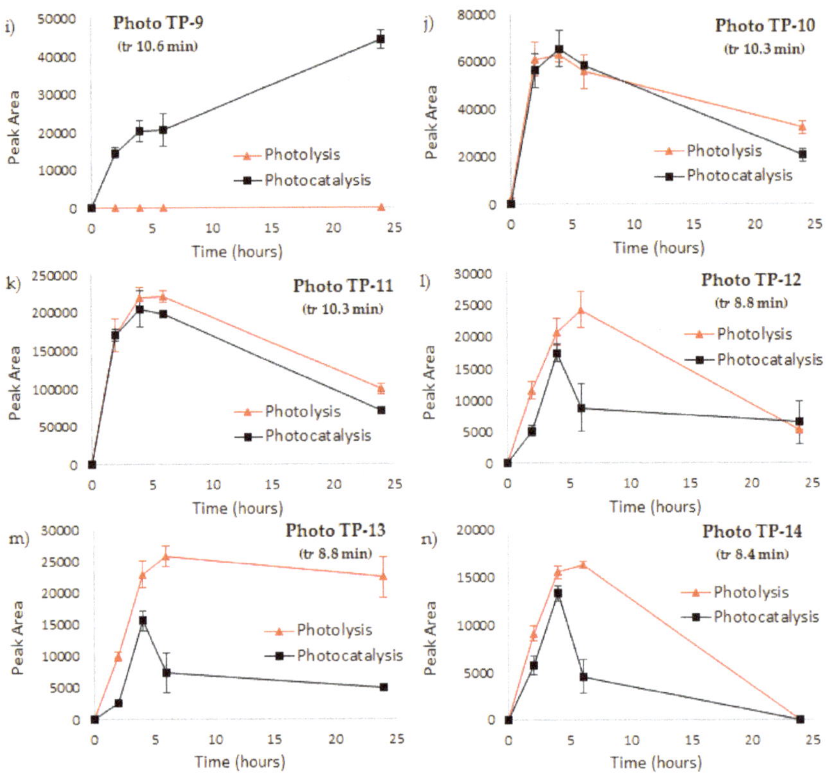

Figure 3. Time profiles of detected Photo TPs during investigated treatments, namely, photolysis (only UV light) and photocatalysis (UV light in the presence of HApTi). Error bars refer to standard deviation of triplicate samples.

The time profiles of the identified Photo TPs (Figure 3) show the average values of the peak area measured during three subsequent cycles of both photolytic and photocatalytic treatments for DCF removal. In each cycle of photocatalysis, the HApTi catalyst was removed from the aqueous solution by means of a centrifugation step at the end of the treatment and then reused for a new subsequent batch of photocatalytic treatment. The results confirm the general trend of identified Photo TPs, for which some transformation products tend to accumulate in the water matrix and some others are transiently formed with a tendency to disappear. Moreover, the results showed that the photocatalytic activity of HApTi catalyst is approximately the same in different subsequent cycles of treatment.

3.2. DCF Degradation: Photolysis versus Photocatalysis with HApTi

Based on the 14 identified Photo TPs, a pathway of DCF degradation in the presence of HApTi was proposed (Figure 2). The principal degradation mechanisms were explained by hydroxylation, oxidation, and decarboxylation reactions. Among these products, six Photo TPs were also identified during photolytic treatment (Figure 3 h, j–n) in the presence of only UV light. Moreover, 22 additional molecular ions (Table 2) showing significant abundance only in the LC-MS chromatograms of photolysis samples by comparing the initial and the last sample (time zero with respect to reaction time 24 h) were detected between 13.3 and 13.6 min (Figure 4) by nontarget screening in positive mode ionization.

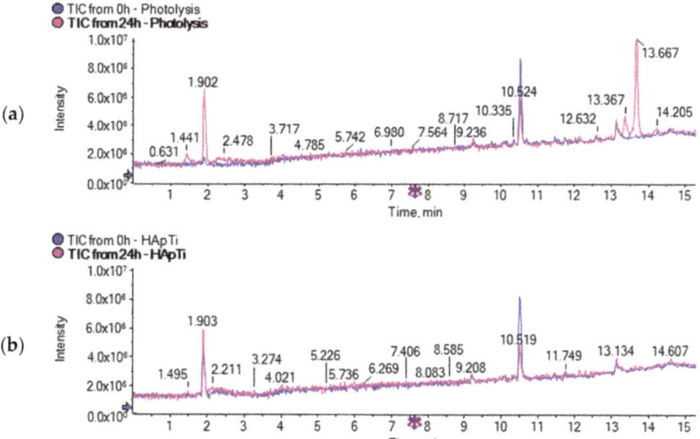

Figure 4. Total ion chromatogram acquired for both (**a**) photolysis and (**b**) photocatalysis in the presence of HApTi comparing time zero with respect to time 24 h.

Table 2. DCF photo-transformation products detected in positive ionization mode in the 24-h photolysis sample.

Compounds	Measured m/z	Retention Time (min)	MS Error (ppm)	Predicted Formula	Formula Finder Score	Profile as Function of Time
Ion-1	230.2471	13.37	1	C14H31NO	39.7	Increase
Ion-2	258.2789	14.18	−1	C16H35NO	40	Increase
Ion-3	352.3052	13.66	0.2	C14H37N7O3	18.1	Increase
Ion-4	379.3046	13.67	0	C21H38N4O2	43.1	Increase
Ion-5	383.2867	13.67	0.1	C17H34N8O2	73.7	Increase
Ion-6	396.3316	13.67	−0.8	C16H41N7O4	25.4	Increase
Ion-7	427.3124	13.66	0.4	C19H38N8O3	67.8	Increase
Ion-8	440.3575	13.67	−0.7	C23H45N5O3	23.2	Increase
Ion-9	449.3258	13.66	−0.4	C23H40N6O3	26.3	Increase
Ion-10	459.4875	13.38	0.2	No formula found	0	Increase
Ion-11	471.3387	13.65	−0.1	C21H42N8O4	33.8	Increase
Ion-12	484.3839	13.68	-1	C25H49N5O4	25.6	Increase
Ion-13	493.352	13.65	−0.6	C22H48N6O4S	72.9	Increase
Ion-14	537.3772	13.64	0.7	C20H48N12O3S	40.5	Increase
Ion-15	559.3899	13.64	1	C21H46N14O4	45.3	Increase
Ion-16	572.4355	13.68	−0.2	C26H49N15	37.8	Increase
Ion-17	577.3903	13.69	0.5	C38H48N4O	91.8	Increase
Ion-18	581.4031	13.63	0.7	C25H48N12O4	14.9	Increase
Ion-19	621.4163	13.68	0.3	C30H48N14O	30.6	Increase
Ion-20	704.5142	13.67	−0.6	No formula found	0	Increase
Ion-21	748.5391	13.67	0.1	No formula found	0	Increase
Ion-22	792.5668	13.66	−0.4	No formula found	0	Increase

An accurate processing of the 20 detected molecular ions by AB-Sciex software allowed us to define for each one a consistent elemental composition based on the exact mass and isotopic pattern. The obtained elemental compositions seem to be mainly nitrogen-containing compounds, but since not enough data are available in the literature about these products, further investigation would be needed in order to identify a relative molecular structure.

These 20 TPs were detected only in samples collected during photolytic treatment; indeed, no evidence of such TPs was found in samples subjected to photocatalytic treatment in the presence of HApTi. This clearly indicates that the photodegradation mechanisms are significantly different when the HAp-based photocatalyst is employed.

It is known that different reactive oxygen species (ROS) are formed with a photocatalyst under light irradiation [30], including •OH, O_{2-}, and h+. Depending on the nature of the catalyst and the experimental conditions, some ROS will prevail over others and will determine the photodegradation mechanism [48]. This affects the nature of the TPs formed during degradation. Regarding the HApTi photocatalyst, some studies report •OH as the main species [49,50]. The present investigation seems to confirm these data, as the identified TPs PhotoTP-1 and TP-2 both have an OH group added to the parent compound. Neither of these products was detected in photolysis experiments (see Figure 3a,b); this confirms a different mechanism that does not involve such ROS.

The TPs identified for the experiment without the catalyst (Figure 3) seem to indicate that UV irradiation induces the breaking of some chemical bonds (i.e., C–Cl) and the formation of some others (i.e., N–H) in the DCF molecule. The identification of the TPs listed in Table 2 could give some further clarifications. This study will be performed in the future as a separate investigation; no further details were added here since the main aim of the present work was to assess the performance of the HApTi photocatalyst in terms of TP toxicity.

3.3. Evaluation of Toxicity of the Treated Water

Figure 5 shows the evolution of the acute toxicity on *D. magna* during photodegradation experiments. DCF solutions (5 mg/L) at the beginning of the experiments exhibit high toxicity on *D. magna* (100% inhibition). It can be seen that the toxicity of the water samples obtained from the photolysis experiment (i.e., no catalyst, Figure 5a) did not vary significantly and remained very high after 24 h of UV irradiation; similar results were observed for a 48-h exposition. When the photocatalyst was employed (Figure 5b), however, the toxicity drastically decreased after 4 h (15% inhibition) and completely disappeared at the end of the experiment (24 h).

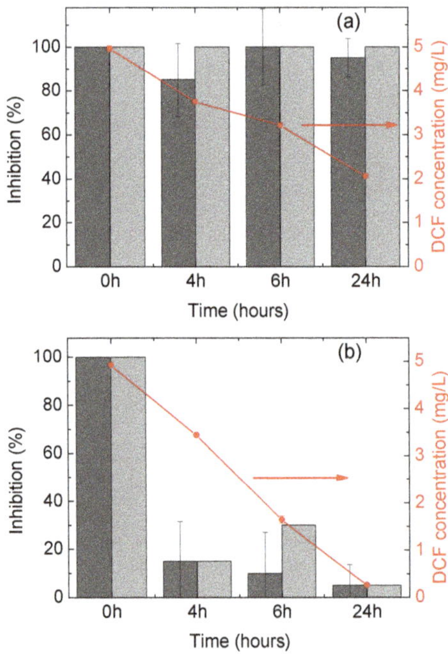

Figure 5. Acute toxicity on *Daphnia magna* during photodegradation of DCF using (**a**) UV light and (**b**) UV light + HApTi. Bars in dark grey refer to 24-h immobilization, while those in pale grey to 48-h immobilization (values to be read on the left axis). Error bars refer to standard deviation of quadruplicate measurements of combined duplicate samples. DCF concentration is reported on the right axis and error bars refer to standard deviation of duplicate samples.

These results indicate that, despite DCF mineralization not being complete, a significant reduction of water toxicity was achieved; this makes photodegradation with the HApTi catalyst a very suitable method for water decontamination.

As opposed to what happened in the test with *D. magna*, in the Toxi-ChromoTest, the original DCF solution did not present any toxicity to the mutant bacteria, which was able to produce the β-galactosidase enzyme and convert the chromogenic substrate at levels similar to negative control (Figure 6). In the photolysis experiment, the toxicity factor increased till 6 h, reaching a maximum of 11% inhibition; it successively decreased. In the presence of the catalyst, the toxicity increased faster, reaching its maximum at 4 h (6.8%) but then decreased, with the final toxicity being lower than that observed in the experiment in the absence of the catalyst. Since the original solution of DCF did not show toxicity, it can be concluded that the toxicity observed in this test for the samples from the photodegradation experiments is exclusively due to the toxicity of the products formed. The difference in toxicity between the two experiments can be related to the different rates at which the TPs are formed and then degraded; such a rate is faster when the HApTi photocatalyst is employed.

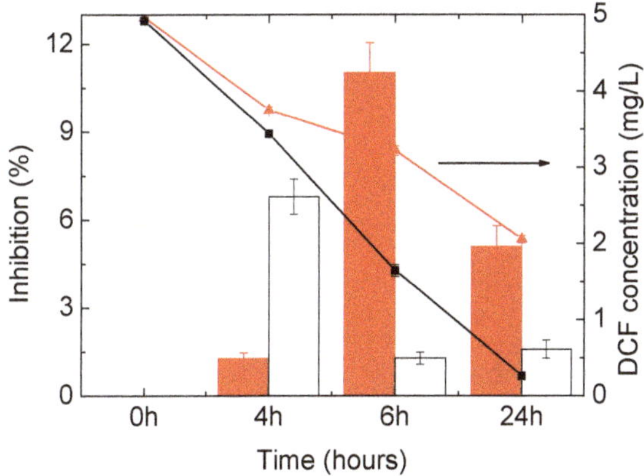

Figure 6. Acute toxicity on mutant bacteria using the Toxi-ChromoTest kit. Red and white bars refer to photolysis and photocatalysis experiments, respectively; the values should be read on the left axis. DCF concentration is reported for both experiments on the right axis—red and black curves respectively. Error bars refer to standard deviation of quadruplicate measurements of combined duplicate samples.

The toxicity trend of *L. sativa* and *S. lycopersicum* exposed to DCF during photodegradation sample (no catalyst) is reported in Figure 7; data for the experiments with the catalyst are not shown since no inhibition at all was observed for the whole photocatalysis process for both seeds. For *L. sativa* (Figure 7a), there was no inhibition of germination for the whole 24-h period of the photolysis experiment. Considering the growth of the germinated seeds, however, a small inhibition of the root growth was observed for the initial DCF sample (0 h). This low toxicity is in agreement with results previously reported for the phytotoxicity of this pharmaceutical on *L. sativa* [1]. The inhibition on root growth increased at 4 h of photolysis; for this time, an inhibition of shoot growth was also registered. The root growth inhibition was lower at 6 and 24 h; however, values still higher than those of the original DCF sample were seen. This is in agreement with literature, since increased phytotoxicity of DCF photolysis transformation products in relation to parent compound was also observed after exposure to sunlight [2]. Moreover, some inhibition was also observed for the shoot growth, which was not registered for the original DCF.

In the case of *S. lycopersicum*, an inhibition of all the analysed parameters was seen with the exposure to the initial DCF sample; also, the inhibition of germination increased during photolysis.

This indicates that exposure to DCF intermediates produced in the absence of photocatalyst was more inhibitory for seed development than the initial DCF solution.

These data show that for both seeds, the intermediates produced by photolysis, i.e., with no photocatalyst, are more toxic than the parent DCF compound; for the photocatalytic experiments, on the other hand, no toxic effects were observed. These data show that photocatalysis using the HApTi catalyst is a more effective way to degrade DCF without generating toxic compounds. Faster DCF removal and faster metabolite transformation may also play a role.

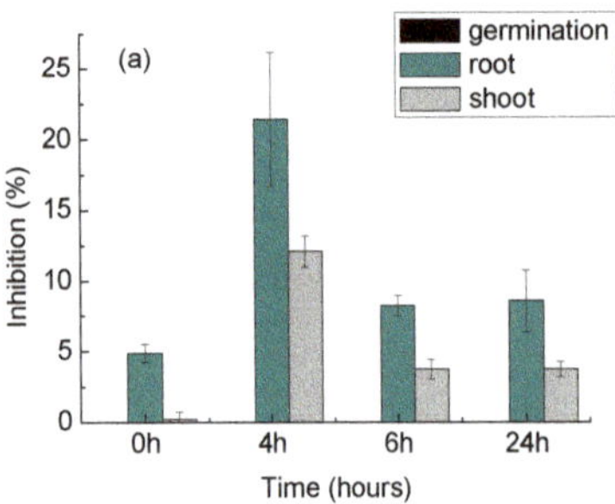
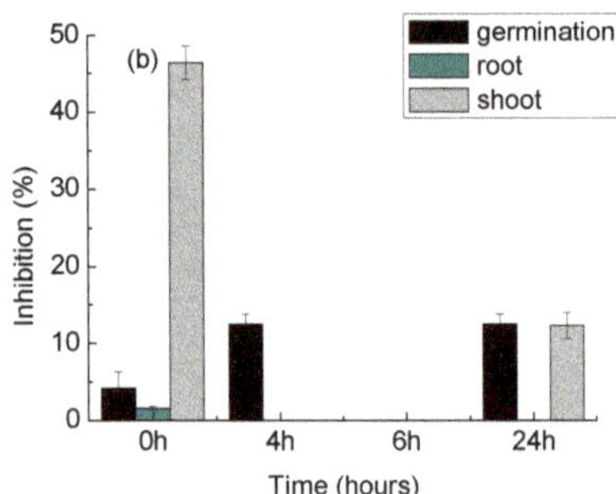

Figure 7. Acute toxicity on (**a**) *Lactuca sativa* and (**b**) *Solanum lycopersicum* during photodegradation of DCF using UV light only (i.e., no catalyst). Experiments with the catalyst showed no toxicity at all (data not shown). Error bars refer to standard deviation of quadruplicate measurements of combined duplicate samples.

4. Conclusions

The present investigation demonstrated the effectiveness of the novel multiphasic hydroxyapatite–TiO$_2$ material for the photocatalytic treatment of diclofenac. Specifically, the novel material gave rise to same transformation products of the photolytic treatment, but the abundance of most of them was minimized at shorter reaction times. It is worth noting that the novel employed catalyst has a size mesh much greater than a conventional TiO$_2$ catalyst and, consequently, its removal at the end of the water treatment process is much easier.

As for the toxicity investigation, it was observed that reduced toxicity of the samples resulted from the photocatalytic experiment in comparison with photolysis and in relation to the original DCF sample. Indeed, the hydroxyapatite–TiO$_2$ material was effective at detoxifying the samples containing DCF and DCF transformation products. It led to a significant decrease in the toxicity of the water samples despite DCF mineralization not being complete. Therefore, such a catalyst represents an interesting and eco-safe technology for DCF degradation.

Author Contributions: C.P. and I.S.M. conceived and designed the experiments; C.P. prepared the photocatalytic material; I.S.M. performed the photodegradation and toxicity experiments; S.M. performed the LC-MS analyses and identification of transformation products. S.M. and I.S.M. wrote the paper. G.M. and P.M.L.C. finalized the paper. G.V. analysed the data; C.C. contributed to finalizing the paper; all authors have read, commented and approved the final manuscript.

Funding: This work was supported by FCT through the project MultiBiorefinery SAICTPAC/0040/2015 (POCI-01-0145-FEDER-016403). The work was performed within the Bilateral Agreement CNR/FCT—Biennial Programme 2015–2016. The work was also partially supported by the project FONTANAPULIA (Fotocatalizzatori NanosTrutturati e RAdiazioNe UV per un'Acqua più PULItA) funded by the EU, Italy, and the Puglia Region through the INNONETWORK call.

Acknowledgments: Irina S. Moreira wishes to acknowledge the research grant from Fundação para a Ciência e Tecnologia (FCT), Portugal (Ref. SFRH/BPD/87251/2012) and Fundo Social Europeu (Programa Operacional Potencial Humano (POPH), Quadro de Referência Estratégico Nacional (QREN)). C.P. would like to thank "Fondazione con il Sud" for financial support (project HapECork, PDr_2015-0243). We would also like to thank the scientific collaboration of CBQF under the FCT project UID/Multi/50016/2013.

Conflicts of Interest: The authors declare no conflict of interest.

References

1. Tran, N.H.; Reinhard, M.; Gin, K.Y.-H. Occurrence and fate of emerging contaminants in municipal wastewater treatment plants from different geographical regions-a review. *Water Res.* **2018**, *133*, 182–207. [CrossRef] [PubMed]

2. Gogoi, A.; Mazumder, P.; Tyagi, V.K.; Tushara Chaminda, G.G.; An, A.K.; Kumar, M. Occurrence and fate of emerging contaminants in water environment: A review. *Ground. Sustain. Dev.* **2018**, *6*, 169–180. [CrossRef]

3. Peake, B.M.; Braund, R.; Tong, A.; Tremblay, L.A. Degradation of pharmaceuticals in wastewater. In *The Life-Cycle of Pharmaceuticals in the Environment*; Elsevier: New York, NY, USA, 2016.

4. Fatta-Kassinos, D.; Vasquez, M.I.; Kümmerer, K. Transformation products of pharmaceuticals in surface waters and wastewater formed during photolysis and advanced oxidation processes—Degradation, elucidation of byproducts and assessment of their biological potency. *Chemosphere* **2011**, *85*, 693–709. [CrossRef] [PubMed]

5. Chen, W.-L.; Cheng, J.-Y.; Lin, X.-Q. Systematic screening and identification of the chlorinated transformation products of aromatic pharmaceuticals and personal care products using high-resolution mass spectrometry. *Sci. Total Environ.* **2018**, *637–638*, 253–263. [CrossRef] [PubMed]

6. Bletsou, A.A.; Jeon, J.; Hollender, J.; Archontaki, E.; Thomaidis, N.S. Targeted and non-targeted liquid chromatography-mass spectrometric workflows for identification of transformation products of emerging pollutants in the aquatic environment. *TrAC Trends Analy. Chem.* **2015**, *66*, 32–44. [CrossRef]

7. Liu, Y.; Wang, L.; Pan, B.; Wang, C.; Bao, S.; Nie, X. Toxic effects of diclofenac on life history parameters and the expression of detoxification-related genes in Daphnia magna. *Aquat. Toxicol.* **2017**, *183*, 104–113. [CrossRef] [PubMed]

8. Välitalo, P.; Massei, R.; Heiskanen, I.; Behnisch, P.; Brack, W.; Tindall, A.J.; Du Pasquier, D.; Küster, E.; Mikola, A.; Schulze, T.; et al. Effect-based assessment of toxicity removal during wastewater treatment. *Water Res.* **2017**, *126*, 153–163. [CrossRef] [PubMed]

9. Van den Brandhof, E.-J.; Montforts, M. Fish embryo toxicity of carbamazepine, diclofenac and metoprolol. *Ecotoxicol. Environ. Saf.* **2010**, *73*, 1862–1866. [CrossRef] [PubMed]

10. Schröder, P.; Helmreich, B.; Škrbić, B.; Carballa, M.; Papa, M.; Pastore, C.; Emre, Z.; Oehmen, A.; Langenhoff, A.; Molinos, M.; et al. Status of hormones and painkillers in wastewater effluents across several European states—Considerations for the EU watch list concerning estradiols and diclofenac. *Environ. Sci. Pollut. Res.* **2016**, *23*, 12835–12866. [CrossRef] [PubMed]

11. Hug, C.; Ulrich, N.; Schulze, T.; Brack, W.; Krauss, M. Identification of novel micropollutants in wastewater by a combination of suspect and nontarget screening. *Environ. Pollut.* **2014**, *184*, 25–32. [CrossRef] [PubMed]

12. Wilkinson, J.; Hooda, P.S.; Barker, J.; Barton, S.; Swinden, J. Occurrence, fate and transformation of emerging contaminants in water: An overarching review of the field. *Environ. Pollut.* **2017**, *231*, 954–970. [CrossRef] [PubMed]

13. Bouju, H.; Nastold, P.; Beck, B.; Hollender, J.; Corvini, P.F.X.; Wintgens, T. Elucidation of biotransformation of diclofenac and 4′hydroxydiclofenac during biological wastewater treatment. *J. Hazard. Mater.* **2016**, *301*, 443–452. [CrossRef] [PubMed]

14. Vieno, N.; Sillanpää, M. Fate of diclofenac in municipal wastewater treatment plant—A review. *Environ. Int.* **2014**, *69*, 28–39. [CrossRef] [PubMed]

15. Salgado, R.; Pereira, V.J.; Carvalho, G.; Soeiro, R.; Gaffney, V.; Almeida, C.; Cardoso, V.V.; Ferreira, E.; Benoliel, M.J.; Ternes, T.A.; et al. Photodegradation kinetics and transformation products of ketoprofen, diclofenac and atenolol in pure water and treated wastewater. *J. Hazard. Mater.* **2013**, *244–245*, 516–527. [CrossRef] [PubMed]

16. Acuña, V.; Ginebreda, A.; Mor, J.R.; Petrovic, M.; Sabater, S.; Sumpter, J.; Barceló, D. Balancing the health benefits and environmental risks of pharmaceuticals: Diclofenac as an example. *Environ. Int.* **2015**, *85*, 327–333. [CrossRef] [PubMed]

17. Zhang, Y.; Geißen, S.-U.; Gal, C. Carbamazepine and diclofenac: Removal in wastewater treatment plants and occurrence in water bodies. *Chemosphere* **2008**, *73*, 1151–1161. [CrossRef] [PubMed]

18. Lonappan, L.; Brar, S.K.; Das, R.K.; Verma, M.; Surampalli, R.Y. Diclofenac and its transformation products: Environmental occurrence and toxicity—A review. *Environ. Int.* **2016**, *96*, 127–138. [CrossRef] [PubMed]

19. Del Moro, G.; Mancini, A.; Mascolo, G.; Di Iaconi, C. Comparison of UV/H$_2$O$_2$ based AOP as an end treatment or integrated with biological degradation for treating landfill leachates. *Chem. Eng. J.* **2013**, *218*, 133–137. [CrossRef]

20. Pagano, M.; Lopez, A.; Volpe, A.; Mascolo, G.; Ciannarella, R. Oxidation of nonionic surfactants by Fenton and H2O2/UV processes. *Environ. Technol.* **2008**, *29*, 423–433. [CrossRef] [PubMed]

21. Kanakaraju, D.; Glass, B.D.; Oelgemöller, M. Advanced oxidation process-mediated removal of pharmaceuticals from water: A review. *J. Environ. Manag.* **2018**, *219*, 189–207. [CrossRef] [PubMed]

22. Miklos, D.B.; Remy, C.; Jekel, M.; Linden, K.G.; Drewes, J.E.; Hübner, U. Evaluation of advanced oxidation processes for water and wastewater treatment—A critical review. *Water Res.* **2018**, *139*, 118–131. [CrossRef] [PubMed]

23. Dewil, R.; Mantzavinos, D.; Poulios, I.; Rodrigo, M.A. New perspectives for Advanced Oxidation Processes. *J. Environ. Manag.* **2017**, *195*, 93–99. [CrossRef] [PubMed]

24. Shaham-Waldmann, N.; Paz, Y. Away from TiO$_2$: A critical minireview on the developing of new photocatalysts for degradation of contaminants in water. *Mater. Sci. Semicond. Process.* **2016**, *42*, 72–80. [CrossRef]

25. Byrne, C.; Subramanian, G.; Pillai, S.C. Recent advances in photocatalysis for environmental applications. *J. Environ. Chem. Eng.* **2018**, *6*, 3531–3555. [CrossRef]

26. Gar Alalm, M.; Tawfik, A.; Ookawara, S. Enhancement of photocatalytic activity of TiO$_2$ by immobilization on activated carbon for degradation of pharmaceuticals. *J. Environ. Chem. Eng.* **2016**, *4*, 1929–1937. [CrossRef]

27. Murgolo, S.; Yargeau, V.; Gerbasi, R.; Visentin, F.; El Habra, N.; Ricco, G.; Lacchetti, I.; Carere, M.; Curri, M.L.; Mascolo, G. A new supported TiO$_2$ film deposited on stainless steel for the photocatalytic degradation of contaminants of emerging concern. *Chem. Eng. J.* **2017**, *318*, 103–111. [CrossRef]

28. Murgolo, S.; Petronella, F.; Ciannarella, R.; Comparelli, R.; Agostiano, A.; Curri, M.L.; Mascolo, G. UV and solar-based photocatalytic degradation of organic pollutants by nano-sized TiO$_2$ grown on carbon nanotubes. *Catal. Today* **2015**, *240*, 114–124. [CrossRef]

29. Comparelli, R.; Cozzoli, P.D.; Curri, M.L.; Agostiano, A.; Mascolo, G.; Lovecchio, G. Photocatalytic degradation of methyl-red by immobilised nanoparticles of TiO$_2$ and ZnO. *Water Sci. Technol.* **2004**, *49*, 183–188. [CrossRef] [PubMed]

30. Piccirillo, C.L.; Castro, P.M. Calcium hydroxyapatite-based photocatalysts for environment remediation: Characteristics, performances and future perspectives. *J. Environ. Manag.* **2017**, *193*, 79–91. [CrossRef] [PubMed]

31. Piccirillo, C.D.C.W.; Pullar, R.C.; Tobaldi, D.M.; Labrincha, J.A.; Parkin, I.P.; Pintado, M.M.; Castro, P.M.L. Calcium phosphate-based materials of natural origin showing photocatlytic activity. *J. Mater. Chem. A* **2013**, *1*, 6452–6461. [CrossRef]

32. Márquez Brazón, E.; Piccirillo, C.; Moreira, I.S.; Castro, P.M.L. Photodegradation of pharmaceutical persistent pollutants using hydroxyapatite-based materials. *J. Environ. Manag.* **2016**, *182*, 486–495. [CrossRef] [PubMed]

33. Poirier-Larabie, S.; Segura, P.A.; Gagnon, C. Degradation of the pharmaceuticals diclofenac and sulfamethoxazole and their transformation products under controlled environmental conditions. *Sci. Total Environ.* **2016**, *557–558*, 257–267. [CrossRef] [PubMed]

34. Deeb, A.A.; Stephan, S.; Schmitz, O.J.; Schmidt, T.C. Suspect screening of micropollutants and their transformation products in advanced wastewater treatment. *Sci. Total Environ.* **2017**, *601–602*, 1247–1253. [CrossRef] [PubMed]

35. Schmidt, S.; Hoffmann, H.; Garbe, L.-A.; Schneider, R.J. Liquid chromatography–tandem mass spectrometry detection of diclofenac and related compounds in water samples. *J. Chromatogr. A* **2018**, *1538*, 112–116. [CrossRef] [PubMed]

36. Detomaso, A.; Mascolo, G.; Lopez, A. Characterization of carbofuran photodegradation by-products by liquid chromatography/hybrid quadrupole time-of-flight mass spectrometry. *Rapid Commun. Mass Spectrom.* **2005**, *19*, 2193–2202. [CrossRef] [PubMed]

37. Piccirillo, C.; Pinto, R.A.; Tobaldi, D.M.; Pullar, R.C.; Labrincha, J.A.; Pintado, M.M.E.; Castro, P.M.L. Light induced antibacterial activity and photocatalytic properties of Ag/Ag3PO4 -based material of marine origin. *J. Photochem. Photobiol. A Chem.* **2015**, *296*, 40–47. [CrossRef]

38. Bessa, V.S.; Moreira, I.S.; Tiritan, M.E.; Castro, P.M.L. Enrichment of bacterial strains for the biodegradation of diclofenac and carbamazepine from activated sludge. *Int. Biodeterior. Biodegr.* **2017**, *120*, 135–142. [CrossRef]

39. Ribeiro, A.R.G.V.M.F.; Maia, A.S.; Carvalho, M.F.; Castro, P.M.L.; Tiritan, M.E. Microbial degradation of pharmaceuticals followed by a simple HPLC-DAD method. *J. Environ. Sci. Health* **2012**, *47*, 2151–2158. [CrossRef] [PubMed]

40. Loos, M. enviMass version 3.5 LC-HRMS trend detection workflow—R package. *Zenodo* **2018**. [CrossRef]

41. OECD. Daphnia acute Immobilisation Test and Reproduction Test. *OECD Guidel. Test. Chem.* **1984**, *202*, 1–16.

42. ISO. *Water Quality: Determination of the Inhibition Mobility of Daphnia Magna STRAUS (Cladocera, Crustacea)—Acute Toxicity Test*; ISO: Geneva, Switzerland, 2013.

43. OECD Guidelines for the Testing of Chemicals 2006. In *Test No. 208: Terrestrial Plant Test: Seedling Emergence and Seedling Growth Test*; Organisation for Economic Co-Operation and Development: Paris, France, 2006.

44. Moreira, I.S.; Bessa, V.S.; Murgolo, S.; Piccirillo, C.; Mascolo, G.; Castro, P.M.L. Biodegradation of Diclofenac by the bacterial strain Labrys portucalensis F11. *Ecotoxicol. Environ. Saf.* **2018**, *152*, 104–113. [CrossRef] [PubMed]

45. Martínez, C.; Canle, L.M.; Fernández, M.I.; Santaballa, J.A.; Faria, J. Aqueous degradation of diclofenac by heterogeneous photocatalysis using nanostructured materials. *Appl. Catal. B Environ.* **2011**, *107*, 110–118. [CrossRef]

46. Agüera, A.P.E.L.A.; Ferrer, I.; Thurman, E.M.; Malato, S.; Fernàndez-Alba, A.R. Application of time-of-flight mass spectrometry to the analysis of phototransformation products of diclofenac in water under natural sunlight. *J. Mass Spectrom.* **2005**, *40*, 908–915. [CrossRef] [PubMed]

47. Calza, P.; Sakkas, V.A.; Medana, C.; Baiocchi, C.; Dimou, A.; Pelizzetti, E.; Albanis, T. Photocatalytic degradation study of diclofenac over aqueous TiO$_2$ suspensions. *Appl. Catal. B Environ.* **2006**, *67*, 197–205. [CrossRef]

48. Abdelhameed, R.M.; Tobaldi, D.M.; Karmaoui, M. Engineering highly effective and stable nanocomposite photocatalyst based on NH2-MIL-125 encirclement with Ag$_3$PO$_4$ nanoparticles. *J. Photochem. Photobiol. A Chem.* **2018**, *351*, 50–58. [CrossRef]

49. Singh, N.; Chakraborty, R.; Gupta, R.K. Mutton bone derived hydroxyapatite supported TiO$_2$ nanoparticles for sustainable photocatalytic applications. *J. Environ. Chem. Eng.* **2018**, *6*, 459–467. [CrossRef]

50. Mohamed, R.M.; Baeissa, E.S. Preparation and characterisation of Pd–TiO$_2$–hydroxyapatite nanoparticles for the photocatalytic degradation of cyanide under visible light. *Appl. Catal. A Gen.* **2013**, *464–465*, 218–224. [CrossRef]

Article

Photoelectrocatalytic Degradation of Paraquat by Pt Loaded TiO$_2$ Nanotubes on Ti Anodes

Levent Özcan [1,*], **Turan Mutlu** [2] **and Sedat Yurdakal** [2,*]

[1] Biyomedikal Mühendisliği Bölümü, Mühendislik Fakültesi, Afyon Kocatepe Üniversitesi, Ahmet Necdet Sezer Kampüsü, 03200 Afyonkarahisar, Turkey
[2] Kimya Bölümü, Fen-Edebiyat Fakültesi, Afyon Kocatepe Üniversitesi, Ahmet Necdet Sezer Kampüsü, 03200 Afyonkarahisar, Turkey; turan.mutlu@redokslab.com
* Correspondence: leventozcan@aku.edu.tr (L.Ö.); sedatyurdakal@gmail.com (S.Y.)

Received: 27 August 2018; Accepted: 11 September 2018; Published: 13 September 2018

Abstract: Nanotube structured TiO$_2$ on Ti surface were prepared in ethylene glycol (Ti/TiO$_2$NTEG) medium by anodic oxidation method with different times and then the plates were calcinated at different temperatures. Non-nanotube structured Ti/TiO$_2$, prepared by thermal oxidation method, and nanotube structured TiO$_2$ on Ti plate in hydrogen fluoride solution were also prepared for comparison. Pt loaded Ti/TiO$_2$NTEG photoanodes were also prepared by cyclic voltammetry method with different cycles and the optimum loaded Pt amount was determined. Photoanodes were characterized by using X-ray Diffraction (XRD), Scanning Electron Microscopy-Energy-Dispersive X-ray Analysis (SEM-EDX), and photocurrent methods. XRD analyses proved that almost all TiO$_2$ is in anatase phase. SEM analyses show that nanotubes and Pt nanoparticles on nanotube surface are dispersed quite homogeneously. The longest nanotubes were obtained in the ethylene glycol medium and the nanotube length increased by increasing applied anodic oxidation time. In addition, a linear correlation between nanotube length and XRD peak intensity was found. Moreover, SEM-EDX and XRD analyses evidence that Pt nanoparticles on nanotube surface are metallic and in cubic structure. Photoelectrocatalytic degradation of paraquat was performed using the prepared photoanodes. Moreover, electrocatalytic and photocatalytic degradations of paraquat were also investigated for comparison, however lower activities were observed. These results evidence that the photoanodes show a significant synergy for photoelectrocatalytic activity.

Keywords: photoelectrocatalysis; TiO$_2$ nanotube; Pt loaded TiO$_2$; paraquat; polar herbicide; degradation

---

1. Introduction

Water pollution caused by organic pollutants (i.e., herbicides, pesticides, pharmaceuticals and care products) is a major problem which should be faced. Some hydrophobic organic compounds could accumulate in oil or similar mediums in which they are very stable and could cause public health problems [1]. Therefore, their elimination needs great efforts. In recent years, the production and application of herbicides and pesticides have been changed; polar and more easily degraded ones have been replaced by nonpolar and stable ones [2].

Paraquat (1,1'-dimethyl-4,4'-bipyridinium dichloride) is a pyridylium herbicide and it is highly used worldwide. Figure 1 shows the chemical structure of paraquat; its polarity and therefore its solubility in water is very high (620 g/L). Although paraquat is prohibited by the European Union, it is still used in almost ninety countries. The high toxicity of paraquat has serious harm to human health and the environment, thus it is important to seek an effective degradation process [3].

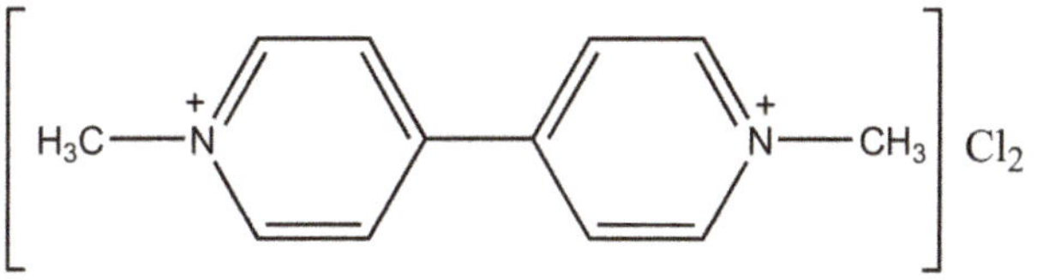

Figure 1. The chemical structure of paraquat (1,1′-dimethyl-4,4′-bipyridinium dichloride).

The chemical oxidation method using compounds such as chlorine, hydrogen peroxide and ozone rarely provides total mineralization of water pollutants. Even though biological oxidation is economical, the presence of poisonous and heat-resistant pollutants in water makes this method insufficient. Resistant organic compounds could be removed by conventional methods such as ultrafiltration and adsorption techniques. However, the main drawback of these methods is that the pollutants pass from one phase to another one without degradation [4].

Photoelectrocatalytic (PEC) techniques could be an effective environmentally friendly alternative for pollutant degradation as it could be performed by using water as solvent, oxygen from air as oxidant and without additional toxic chemicals [5]. TiO_2 is the most used photocatalyst because of its high photocatalytic (PC) activity, biological and chemical stability, low cost, non-toxic formation and non-photocorrosive nature [6–8]. Although the PC process, in which a semiconductor is used as the catalyst and ultraviolet (UV) or ultraviolet-visible (UV-Vis) irradiation as an energy source is a common method for elimination of harmful compounds, their application is limited due to their low quantum yield caused mainly by high recombination rate [4]. Photoelectrocatalysis, i.e., the combination of heterogeneous photocatalysis with electrocatalysis, is an effective tool for hindering the photogenerated pairs recombination and consequently for improving the yield of reactions [8,9]. This method involves the application of an external potential bias to a thin TiO_2 layer deposited on a conductive support [8,10–12]. In other words, under suitable irradiation (i.e., UV or UV-Vis), the used bias allows the electrons to migrate across the electrode and the separation of electron (e^-) and hole (h^+) pairs is improved, thus increasing the probability of reaction occurring at the working electrode surface.

Nowadays, nanostructured TiO_2 is frequently used in PEC studies in many different areas such as degradation of harmful compounds [13,14], water splitting reactions [15,16] or partial oxidations [9,17,18] due to its high surface area and good electron transport characteristics [19].

The first synthesis of TiO_2 by anodic oxidation was investigated with basic peroxide and chromic acid treatment by Assefpour-Dezfuly and coworkers [20]. Zwilling and coworkers [21] investigated nanoporous titanium dioxide in fluoride-containing electrolyte, pioneering a big advancement in this area over the last two decades. Many researchers working in this area have made great efforts to find the optimum electrolyte conditions and experiment parameters to obtain high quality and self-assembled titanium dioxide nanotube array. Gong et al. [22] prepared self-assembled TiO_2 nanotubes with the anodization of Ti plate in H_2O/HF electrolyte medium at room temperature, even though the nanotube length was limited to a few hundred nanometers. Several micrometer length TiO_2 nanotubes have been prepared in neutral electrolytes containing fluoride ions such as Na_2SO_4/NaF or $(NH_4)_2SO_4/NH_4F$ [23,24]. During the anodization process, undulatory and ring shapes were observed in the nanotube TiO_2 walls obtained by current fluctuation. In subsequent works, anodization was carried out in organic electrolytes such as formamide, dimethylsulfoxide, ethylene glycol (EG) or diethylene glycol containing fluoride to produce smooth and several-hundred-micrometer length nanotube TiO_2 [25,26].

The loading of noble metal nanoparticles such as Au [27], Ag [28], Pd [29] and Pt [30] on the TiO_2 nanotube surface is an effective tool to increase its (photo)catalytic properties. It is evidenced that modification of photocatalysts with the noble metals improves the PC and PEC activities by preventing the recombination rate of the electron–hole pairs [31].

In a previous study, we investigated the PEC synthesis of *p*-anisaldehyde from 4-metoxybenzyl alcohol in water, under UV irradiation [9] by using Ti/TiO$_2$ photoanodes prepared both by thermal oxidation and dip-coating methods. The photoanodes prepared by thermal oxidation showed better aldehyde selectivity and activity performance than dip-coated ones.

For the first time, we reported the selective PEC oxidation of 5-(hydroxymethyl)-2-furaldehyde to 2,5-furandicarbaldehyde by nanotube structured TiO$_2$ on Ti layer as photoanodes, prepared by anodic oxidation method in HF medium [18]. In addition, the photoanodes were also platinized by photoreduction method and both product selectivity and PEC activity significantly increased.

Althought some PC paraquat degradation works have been reported [1,32,33], in the present work, PEC degradation of paraquat was investigated for the first time. Highly active nanotube structured TiO$_2$ on Ti layers as photoanodes was prepared by anodic oxidation method in EG medium and used in paraquat degradation. The effects of anodic oxidation time on nanotube length and the amount of Pt loading by cyclic voltammetry were also investigated. Photoanodes prepared by thermal oxidation (Ti/TiO$_2$-500) [9] and anodic oxidation in HF medium (Ti/TiO$_2$-NTHF-X) [18] were used for comparison.

2. Materials and Methods

2.1. Preparation of Ti Plates

Ti plates (5.0 cm × 8.0 cm × 0.10 cm) were used for the photoanode preparation. The plates were grinded (800, 1000, 1200 and 1500 grids of emery papers, respectively) to smooth the surface, and then sonicated in acetone, ethanol and water (10 min for each solvent). The Ti plates were chemically cleaned in a solution medium containing 4% HF, 31% HNO$_3$ and 65% H$_2$O for 30 s. Then, the Ti plates were again sonicated in water for 10 min and dried at room temperature.

2.2. Preparation of Thermally Oxidized Photoanode (Ti/TiO$_2$)

The photoanode was produced by oxidizing the Ti plate surfaces by means of thermal annealing in air (temperature increasing rate: 3 °C/min) at 500 °C for 3 h in an oven (PLF-110/10 model, Protherm Furnaces, Ankara, Turkey). The obtained sample was labeled as Ti/TiO$_2$-500, where 500 indicates the thermal treatment temperature expressed in °C. Figure S1 shows the photo of Ti/TiO$_2$-500 photoanode.

2.3. Preparation of Nanotube Structured TiO$_2$ Photoanodes by Anodic Oxidation in Hydrogen Fluoride Solution (Ti/TiO$_2$NTHF-X)

Ti plate immersed in an aqueous solution containing 0.15 M HF [18] were subjected to an anodic oxidation process by applying for different times a 20 V potential with a direct current (DC) power supply device (Meili, MCH-305D-II model, MCH Instrument, Shenzhen, China). Thus, nanotube structured amorphous TiO$_2$ was formed on the Ti plate surface. Figure S2 shows experimental setup system used for the anodic oxidation. In this system, carbon felt electrode was used as the cathode. The carbon felt was immersed in a 1 M HNO$_3$ aqueous solution for one day to make its surface hydrophilic. In addition, the process was carried out at 200 rpm and at 18 °C (±2 °C), in a high-density polyethylene container, instead of glass, because of the use of fluoride containing solutions.

Figure S3 shows the electrodes prepared by anodic oxidation and thermally treated at different temperatures (400–700 °C, temperature increasing rate: 3 °C/min). These photoanodes were named as Ti/TiO$_2$NTHF-X-Y, where "X" refers to the anodic oxidation time (in hours) and "Y" the calcination temperature.

2.4. Preparation of Nanotube Structured TiO$_2$ Photoanodes by Anodic Oxidation in Ethylene Glycol Solution (Ti/TiO$_2$NTEG-X)

The solution used for the anodic oxidation was prepared by dissolving NH$_4$F in a concentration of 0.3% (*w*/*w*) in a solution containing 2% (*v*/*v*) water and 98% (*v*/*v*) EG [34]. Initially, a 60 V potential was used [34], as the nanotubes exfoliated the voltage was decreased to 40 V. The formed TiO_2 nanotubes were in amorphous phase. To transform them to crystalline one, the electrodes were calcined at 450, 500 or 550 °C (temperature increasing rate: 3 °C/min). The prepared electrodes were named as Ti/TiO_2-NTEG-X-Y, where "X" refers to the applied voltage time (in hours) for anodic oxidation in EG and "Y" the heat treatment temperature (°C). The photographs of the photoanodes prepared by this method are shown in Figure S4.

2.5. Pt Loading on Ti/TiO₂NTEG-3h-500 Photoanodes

Pt loading on Ti/TiO_2NTEG-3h-500 electrode was carried out by using a cyclic voltammetry (CV) technique with an aqueous solution of 0.25 mM $H_2PtCl_6 \cdot 6H_2O$ and 25 mM H_2SO_4. The Pt^{4+} ions in solution were electrochemically loaded on the TiO_2 nanotube surface with different cycle numbers (voltage range: -0.4–$+0.5$ V, scan rate: 10 mV/s). Ag/AgCl (3M KCl) electrode was used as the reference electrode and a Ti plate (5.0 cm $\times$ 8.0 cm $\times$ 0.10 cm) was used as the counter electrode. Figure S5 shows voltammograms obtained during Pt nanoparticle loading on Ti/TiO_2NTEG-3h-500 electrode by CV until 4 cycles. The peak at ca. -0.35 V decreased by increasing the cycle number probably due to the decrease of Pt^{4+} ions concentration in solution.

2.6. Photoanode Characterization

X-ray diffraction (XRD) patterns were recorded by a Shimadzu (XRD-6000 model) diffractometer (Shimadzu Corporation, Tokyo, Japan) by using a Cu Kα radiation (1.544 A°) and a 2θ scan rate of 1.281 degree/min^{-1}.

Scanning electron microscope (SEM) images were obtained by using a FEI microscope (NanoSEM 650 model, FEI Company, Hillsboro, OR, USA). SEM images were obtained by using TLD detector. EDX spectra were taken with the EDX detector available in the SEM system. In addition SEM images of Figures 4–6 and S6 were obtained by another SEM instrument (Carl Zeiss ULTRA Plus, Germany).

The performances of the photoanodes were determined chronoamperometrically under UV irradiation in the same electrolyte medium used for PEC paraquat degradation.

2.7. Photo-, Electro- and Photoelectro-Reactivity Set up and Procedure

PEC and EC experiments of paraquat degradation were carried out with a three-electrode electrochemical system connected to a computer-controlled potentiostat-galvanostat (Ivium, CompactStat model, Ivium Technologies, Eindhoven, The Netherlands) device. In this system, Ti/TiO_2 photoanodes, Pt electrode (gauze) and Ag/AgCl (3 M KCl) electrodes were used as working, counter and reference electrodes, respectively. The sizes of the photoanode immersed part were 5.0 cm $\times$ 6.0 cm $\times$ 0.10 cm. In the PEC and PC experiments, 6 UV fluorescent lamps (8 W) with a maximum wavelength of 365 nm were used as the irradiation source. The PEC and EC experiments setup system is shown in Figure S7. The distance between the photoanodes and the light source is 7 cm, and the light intensity at this distance from the photoanode surface is about 13 W/m^2. This value was determined by a radiometer (DO9721, Deltaohm, Caselle di Selvazzano, Italy) with a probe measuring 315–400 nm.

Experiments were carried out in environmentally friendly conditions using water as the solvent and oxygen in the air as the oxidant. The solution was in contact with atmospheric air, therefore the concentration of dissolved oxygen was always assumed to be in equilibrium with oxygen of the atmosphere.

The initial concentration of paraquat was 37.4 µM and the concentration of Na_2SO_4 used as electrolyte was 50 mM. A Pyrex beaker of 150 mL was used as the reactor for degradation of 150 mL of paraquat solution.

Before switching the lamp on, the solution was stirred for 10 min at room temperature under dark condition without using any voltage to reach the thermodynamic equilibrium. During the runs, the solution was maintained at ca. 25 °C.

2.8. Analytical Techniques

Samples were withdrawn at fixed time intervals (0, 30, 60, 120 and 180 min) and analyzed by UV-Vis spectrophotometer (UV-1700 model, Shimadzu Corporation, Tokyo, Japan). A linear calibration graph was obtained from the absorbance values of standard solutions of paraquat. The conversion values (%) were calculated as follows:

$$\% \text{ Conversion} = [(\text{concentration of reacted paraquat})/(\text{initial concentration of paraquat})] \times 100$$

All used chemicals were purchased from Sigma Aldrich (Darmstadt, Germany) with a purity >98.0%.

3. Results and Discussion

3.1. Characterization Results

Figures 2 and 3 show XRD patterns of nanotube TiO_2 supported photoanodes. XRD patterns of pristine Ti layer are also given as comparison. The figures show peaks at $2\theta = 25.58°$, $38.08°$, $48.08°$ and $54.58°$ that belong to anatase phase; peaks at $2\theta = 27.5°$, $36.5°$, $41.0°$, $54.1°$ and $56.5°$ that belong to rutile phase [35]; and peaks at $2\theta = 34.95°$, $38.25°$, $40.05°$ and $52.90°$ that belong to metallic Ti [18]. In addition, the peak values at $2\theta = 39.8°$, $46.2°$ and $67.5°$ refer to metallic platinum [18].

Figure 2 shows XRD patterns of Ti/TiO_2NTEG-3h-450, Ti/TiO_2NTEG-3h-500-Pt-25cycles and Ti/TiO_2NTEG-3h-550 prepared in the EG medium and calcined at different temperatures (450–550 °C). The prepared photoanodes contain bands of TiO_2 mainly in anatase phase and of metallic Ti, whereas only traces of rutile were found. It can be noticed that the peak intensity of Ti decreases by increasing the calcination temperature, while the peak intensity of anatase increase. The most thermodynamically stable phase of TiO_2 is rutile [36], therefore its peak intensity increases by increasing the heat treatment temperature.

Pt was not observed for the sample Ti/TiO_2NTEG-3h-500-Pt-4cycles since Pt amount is not enough to be determined by XRD analysis. On the contrary, XRD analysis of Ti/TiO_2NTEG-3h-500-Pt-25cycles sample show that the loaded Pt is in metallic form (Figure 2 and Figure S8).

Figure 3 shows the XRD patterns of nanotube structure of TiO_2 on Ti prepared in the EG medium and for different anodic oxidation times (1, 3 and 6 h), and then calcined at 500 °C. XRD of the Ti/TiO_2-NTHF-6h-500 electrode, prepared in the HF solution, was added for comparison purposes. TiO_2 in all photoanodes is mainly in anatase with trace amount of rutile phase. The XRD peak intensities of anatase phase significantly increased by increasing anodic oxidation time used for photoanode preparation. Since all these photoanodes were calcined at the same temperature (500 °C), the crystallization degree of anatase phase of the electrodes is similar. Moreover, the increase in peak intensity of anatase phase (peak area values in Table 1) could be related to the amount of TiO_2 on the Ti surface, and thus to the length of the nanotubes. On the contrary, by increase the nanotube lengths, the XRD peaks of metallic Ti decrease. The XRD peak intensity of the electrode prepared by anodic oxidation method in the HF solution (Ti/TiO_2NTHF-6h-500) is very low compared to the others prepared in EG, indicating that the length of the TiO_2 nanotubes of Ti/TiO_2NTHF-6h-500 is also shorter.

Table 1 reports the crystal phases of the photoanodes, the peak areas (101) of anatase and rutile phases, and the primary particle sizes determined from the Scherrer equation. The primary particle sizes of the anatase phase of all the electrodes are close to each other (ca. 35 nm). This value is approximately the same as the wall thickness of TiO_2 nanotubes (see Table 2). Since the rutile peak areas are very low, their results are not reliable enough (range from 23 to 43 nm). The primary particle size of metallic Pt of Ti/TiO_2NTEG-3h-500-Pt-25cycles electrode is 28 nm.

Table 1. Crystal phase, peak areas (101) of anatase and rutile phases, and primary particle sizes of the prepared anodes.

Electrode	Crystal Phase	Peak Area of Anatase (101)	Primary Particle Size (nm) of Anatase	Primary Particle Size (nm) of Rutile	Primary Particle Size (nm) of Pt
Ti/TiO$_2$NTEG-3h-450	A	423	37		
Ti/TiO$_2$NTEG-3h-500-Pt-25cycles	A + R	450	33		28
Ti/TiO$_2$NTEG-3h-550	A + R	477	35	43	
Ti/TiO$_2$NTEG-1h-500	A + R	194	35	22	
Ti/TiO$_2$NTEG-6h-500	A + R	558	38	37	
Ti/TiO$_2$NTHF-6h-500	A + R	48	32	23	

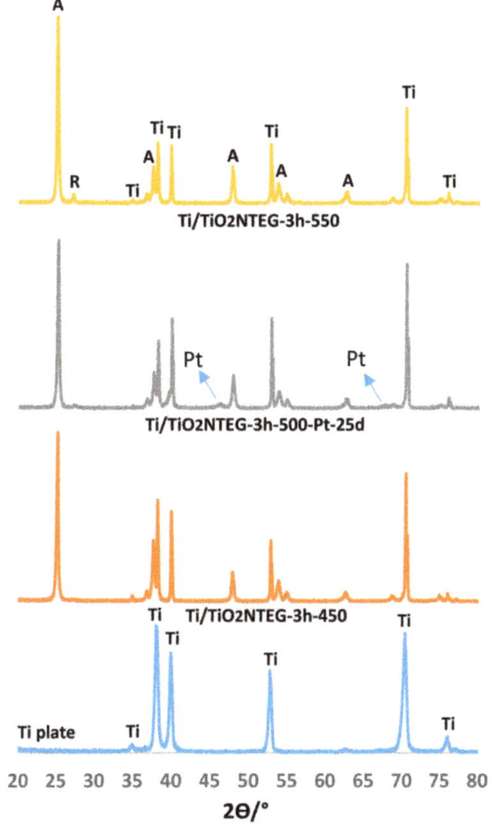

Figure 2. XRD patterns of the Ti/TiO$_2$-NTEG-3h-Y photoanodes.

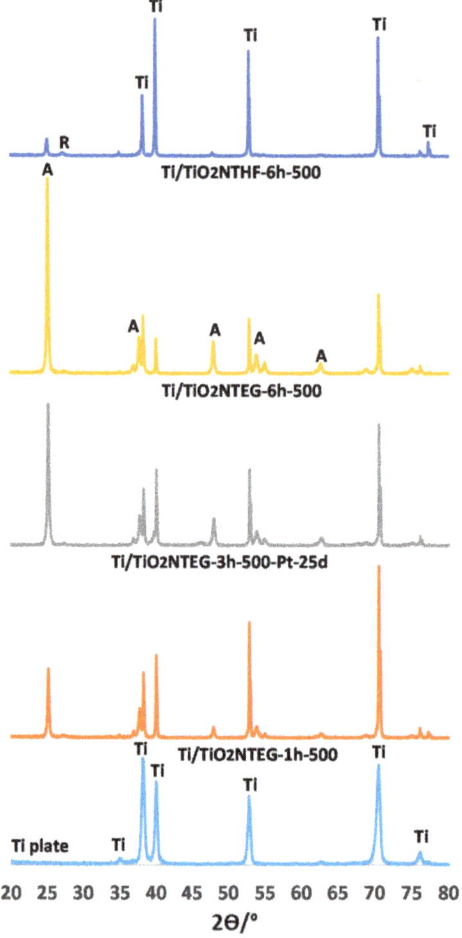

Figure 3. XRD patterns of the Ti/TiO$_2$-NTEG-Xh-500 photoanodes.

According to SEM image (Figure S9) of Ti/TiO$_2$-500, TiO$_2$ on the surface of the Ti plate is in the form of a slightly rough film [18].

SEM image of Ti/TiO$_2$NTHF-1h-500 is shown in Figure S10. The nanotubes prepared by this method showed an independent formation with each other [18]. In other words, there are spaces between the nanotubes and each nanotube does not have a common wall. The values of tube length, wall thickness and tube diameter of nanotubes are reported in Table 2. The nanotubes are distributed homogeneously on the surface, the thickness of the nanotube wall is ca. 14 nm and the diameter of the tube hole is ca. 90 nm.

Table 2. Average wall thickness and internal diameters of nanotubes on the anodes evaluated by SEM imagines.

Electrode	Wall Thickness (nm)	Internal Diameter (nm)	Tube Length (μm)	Pt Nanoparticle Diameter (nm)
Ti/TiO$_2$NTEG-1h-500	35	43	1.7	
Ti/TiO$_2$NTEG-2h-500	30	47		
Ti/TiO$_2$NTEG-3h-500	35	60		
Ti/TiO$_2$NTEG-4h-500	35	47	9.8	
Ti/TiO$_2$NTEG-6h-500	20	80	11	
Ti/TiO$_2$NTHF-6h-500	14	90		
Ti/TiO$_2$NTHF-6h-650	30	75	1.0	
Ti/TiO$_2$NTEG-3h-500-Pt-4cycles	35	60		80
Ti/TiO$_2$NTEG-3h-500-Pt-25cycles	40	60		150

SEM images of Ti/TiO$_2$NTHF-6h-650 are shown in Figure 4 (and Figure S6). It can be noticed that the nanotubes cover only some part of the surface. This aspect is probably due to the higher calcination temperature. Indeed, in Figure 4a, it is possible to distinguish two zones, a clearer one in high ground which refers to TiO$_2$ in nanotubes (Figure 4b) and a darker one that can be attributed to a well crystallized TiO$_2$ in layer (Figure 4c) with particle sizes of ca. 200 nm. In addition, its wall thickness is higher and internal diameter lesser than Ti/TiO$_2$NTHF-6h-500. Total thickness of both values of both plates are almost the same (ca. 105 nm).

A cross-sectional SEM image of the Ti/TiO$_2$NTHF-6h-650 (Figure 4d) allowed determining the nanotube length (ca. 1 μm).

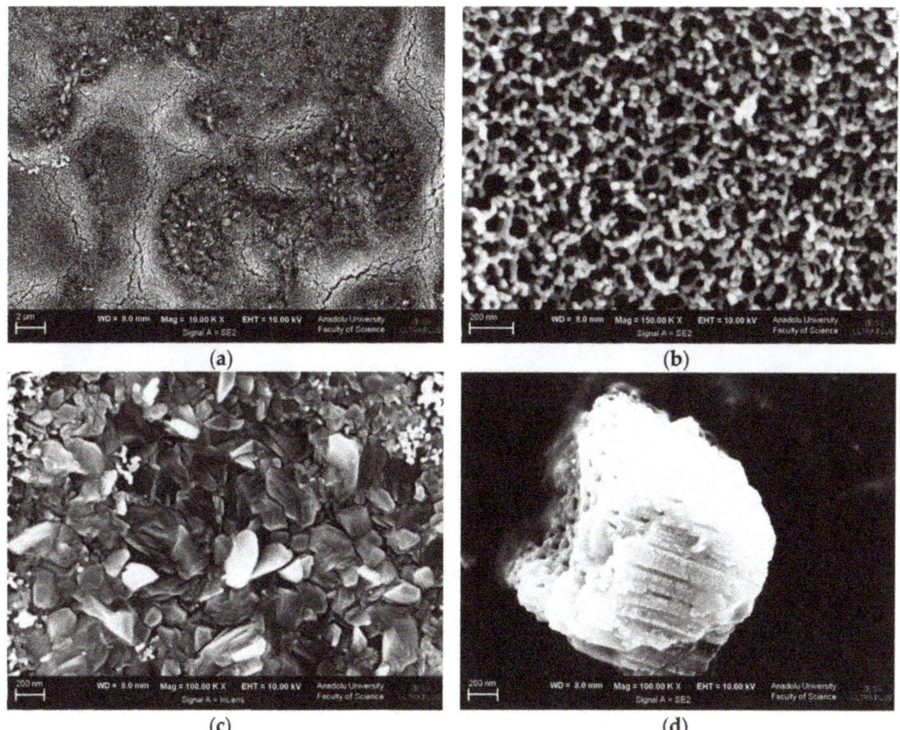

Figure 4. SEM images of Ti/TiO$_2$NTHF-6h-650 photoanode: (**a**) top view (magnification: 10,000×), (**b**) top view (magnification: 150,000×), (**c**) top view (magnification: 100,000×) and (**d**) cross-section view (magnification: 100,000×).

Figure 5 shows SEM images of Ti/TiO$_2$NTEG-1h-500 electrode. Figure 5a indicates a good distribution of the nanotubes on the surface, whereas Figure 5b shows the cross-section image of the nanotubes. It was possible to determine the wall thickness of the nanotubes (35 nm), the diameter of the tube hole (43 nm), and the average nanotube length (1.7 μm).

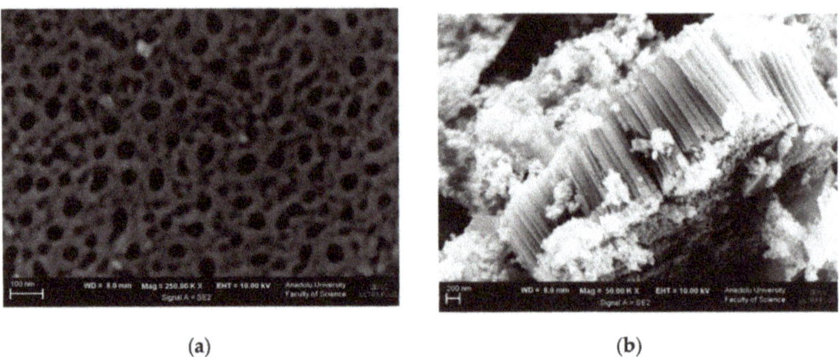

(a) (b)

Figure 5. SEM images of Ti/TiO$_2$NTEG-1h-500 photoanode: (**a**) top view (magnification: 250,000×) and (**b**) cross-section view (magnification: 50,000×).

SEM images of the Ti/TiO$_2$NTEG-2h-500 photoanode are shown in Figure 6. Figure 6a,b show the same image with different magnifications, while Figure 5c shows the cross-section of the material. The nanotubes are distributed quite homogeneously on the surface of the Ti sheet. It can be noticed (Figure 6c) that the nanotube formation occurs even at the bottom of the tubes. However, the inner diameter of the nanotubes in the bottom part is narrower than the top. This result shows that nanotubes form from outside to inside.

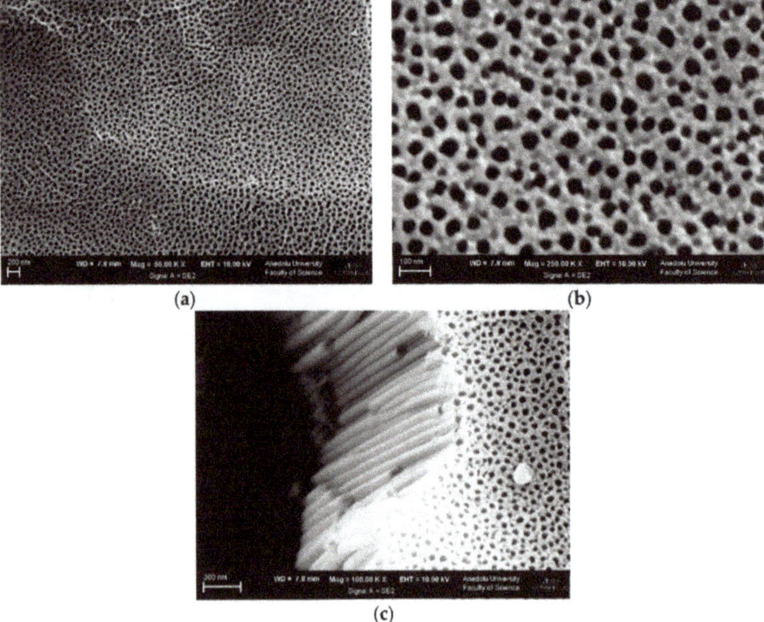

(a) (b)

(c)

Figure 6. SEM image of Ti/TiO$_2$NTEG-2h-500 photoanode: (**a**) top view (magnification: 50,000×), (**b**) top view (magnification: 250,000×) and (**c**) cross-section view (magnification: 100,000×).

Figure 7 (and Figure S11) shows SEM images of the bottom, top and cross-section of Ti/TiO$_2$NTEG-4h-500 photoanode. Ti surface is completely covered with nanotubes, as shown in Figure 7a. Figure 7a,b shows a quite homogeneous distribution of the nanotubes. Figure 7c,d and Figure S11 are different magnified cross–sections of Ti/TiO$_2$NTEG-4h-500 photoanode. The length of the nanotube is about 9.8 µm, and this size extends over a wide range (Figure S11). This nanotube length is greater than the Ti/TiO$_2$NTEG-1h-500 photoanode. Some nanotubes zones are covered by TiO$_2$ sheets (see Figure 7d). The nanotube wall thickness and hole diameter of tube values are close to the Ti/TiO$_2$NTEG-1h-500 electrode (ca. 35 and ca. 47 nm, respectively). However, the wall thickness is thicker at the bottom of the tube, while the hole diameter appears to be smaller (~38 and ~27 nm, respectively, Table 2).

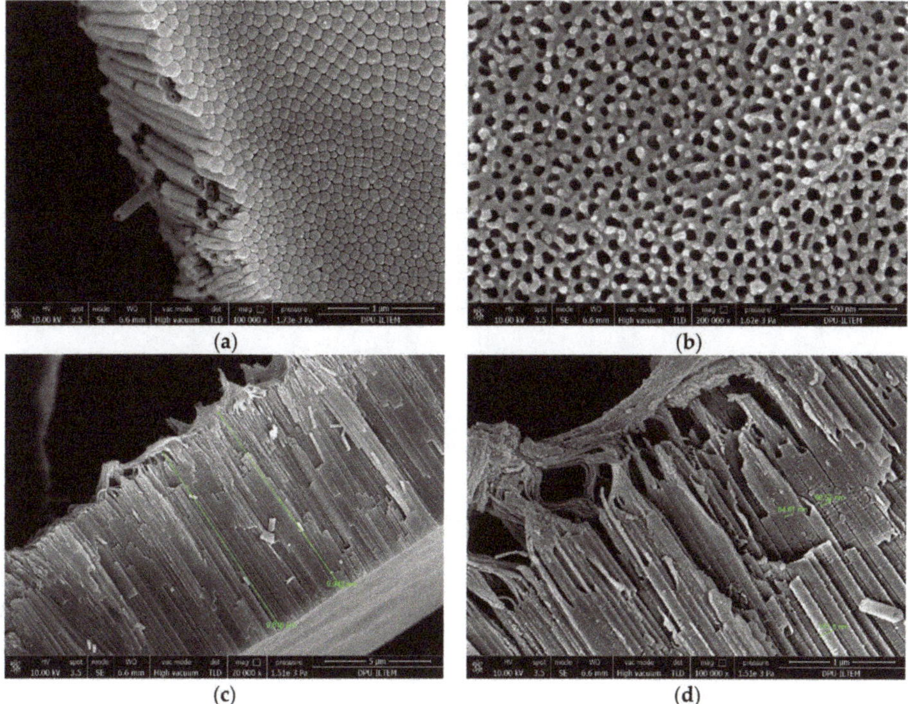

Figure 7. SEM images of Ti/TiO$_2$NTEG-4h-500 photoanode: (**a**) bottom view (magnification: 100,000×); (**b**) top view (magnification: 200,000×); (**c**) cross–section view (magnification: 20,000×); and (**d**) cross–section view (magnification: 100,000×).

Figure 8 shows some SEM images of Ti/TiO$_2$NTEG-6h-500 photoanode from bottom (Figure 8a and Figure S12b), cross–section (Figure 7b and Figure S12a) and top view (Figure 8c,d). From observation of Figure 8b, it is possible to determine that the nanotube length of this electrode is the longest one (ca. 11 µm). The data in Table 2 show that there is a linear relationship between the anodic oxidation time for the preparation and the nanotube length. Similar relationship was also observed from the peak intensities in XRD analyses. Generally, nanotubes are uniformly distributed on the layer surface (Figure 8c), but there are also some regions covered by TiO$_2$ sheets (Figure 8d). This behavior was found for long anodic oxidation times. The wall thickness of the nanotubes for this sample is thinner than the others (ca. 20 nm) and the inner hole diameter is larger (~80 nm, Table 2).

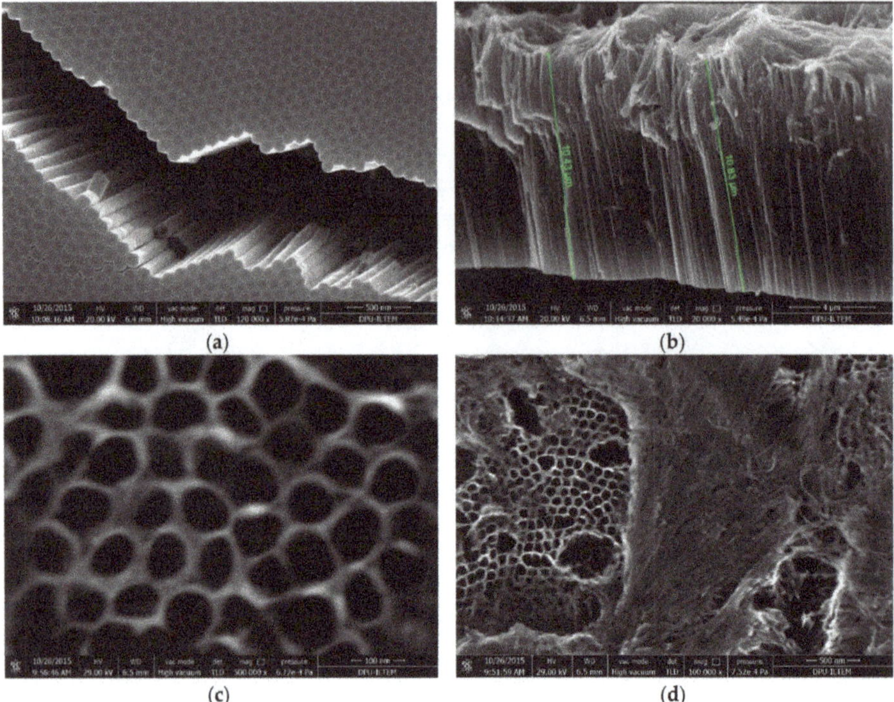

Figure 8. SEM images of Ti/TiO$_2$NTEG-6h-500 photoanode: (**a**) bottom view (magnification: 120,000×); (**b**) cross–section view (magnification: 20,000×); (**c**) top view (magnification: 500,000×); and (**d**) top view (magnification: 100,000×).

The SEM images of both Ti/TiO$_2$NTEG-3h-500 and the corresponding Pt loaded anodes are presented in Figure 9. It is possible to notice the platinum nanoparticle on the external surface of the nanotubes. EDX analysis evidenced that the white nanoparticles belong to Pt. Both TiO$_2$ nanotubes and Pt nanoparticles located outside of the pores are quite homogeneously distributed. The nanotube wall thickness and inner hole diameter of these electrodes are 35 and 60 nm, respectively (Table 2). The average diameter of the Pt nanoparticles is about 80 nm.

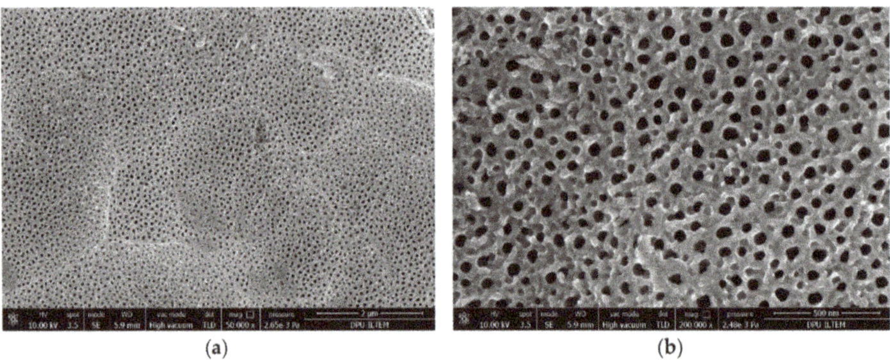

Figure 9. *Cont.*

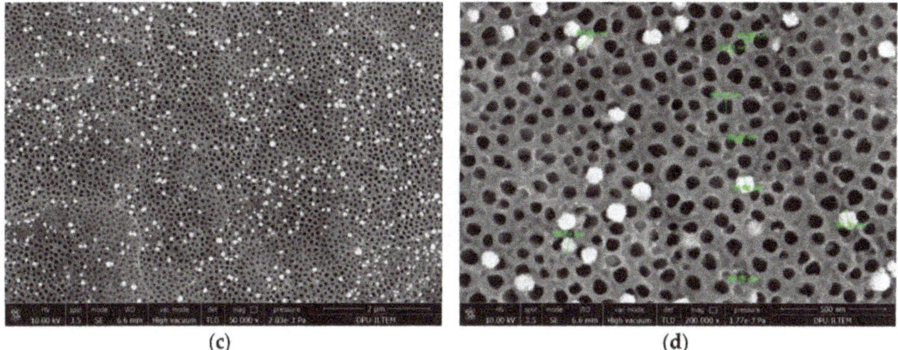

(c)   (d)

Figure 9. SEM images of photoanodes: Ti/TiO$_2$NTEG-3h-500 (**a,b**); and Ti/TiO$_2$NTEG-3h-500-Pt-4cycles (**c,d**). Magnifications: 50,000× (**a,c**); and 200,000× (**b,d**).

Figure 10 shows SEM images of Ti/TiO$_2$NTEG-3h-500-Pt-25cycles photoanode. Although the TiO$_2$ nanotube morphology for this electrode is almost the same of that of Ti/TiO$_2$NTEG-3h-500-Pt-4cycles, Pt nanoparticles are quite different in dimensions (ca. twice). In addition, Pt nanoparticle sizes are quite different from each other and they have a heterogeneous distribution in this electrode. This feature indicates that, by increasing the number of cycles, the initially formed small nanoparticles act as crystallization nuclei which grow without the creation of new particles.

The Pt nanoparticle diameter on the surface of Ti/TiO$_2$NTEG-3h-500-Pt-25cycles electrode is about 150 nm (Table 2). It is clear from SEM image that Pt nanoparticles are in cubic form. In our previous study, TiO$_2$ nanotube surfaces were loaded with Pt by photoreduction method and much smaller and heterogeneously distributed Pt nanoparticles were obtained [18].

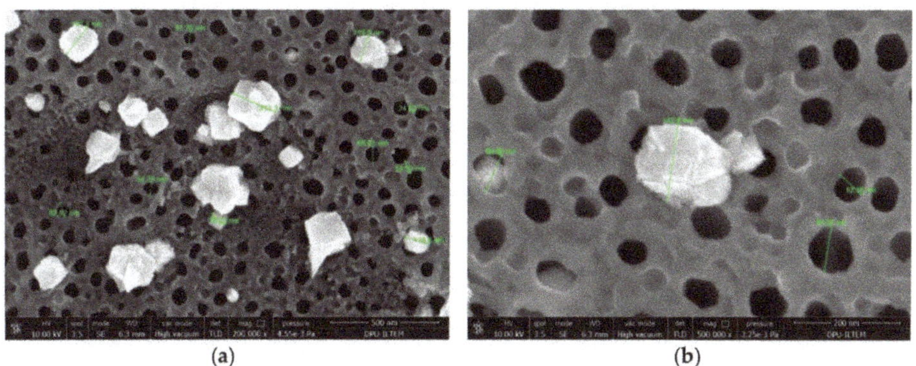

(a)   (b)

Figure 10. SEM images of Ti/TiO$_2$NTEG-3h-500-Pt-25cycles photoanode at magnification: 200,000× (**a**); and 500,000× (**b**).

Figure 11 shows the photocurrent values of electrodes, prepared by different methods, in the same medium in which PEC paraquat degradation reactions were performed. The lowest photocurrent value is attributed to the Ti/TiO$_2$-500 photoanode obtained by the thermal oxidation method with the least effective surface area. The photocurrent values of the electrode prepared in HF medium is a little higher than that of Ti/TiO$_2$-500 because of its higher effective surface area. Therefore, the electrode with very long nanotube (Ti/TiO$_2$NTEG-3h-500) showed the highest photocurrent values, as expected. However, Pt nanoparticles loaded to the Ti/TiO$_2$NTEG-3h-500 surface reduced the absorbed amount of light on TiO$_2$ surface, resulting in a decrease in the photocurrent value.

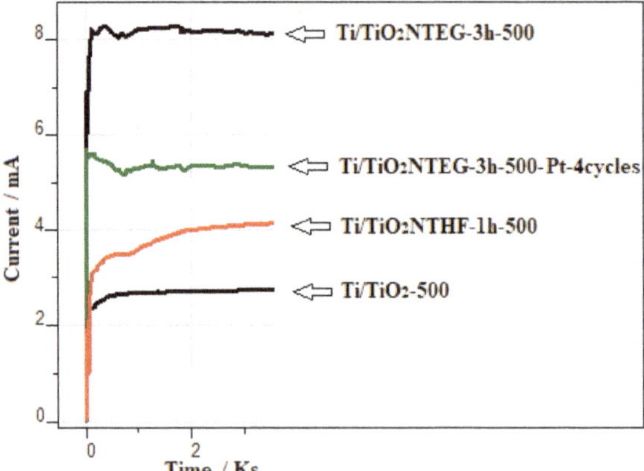

Figure 11. The photocurrent values of electrodes, prepared by different methods, in the same medium in which PEC paraquat degradation reactions were performed.

Figure 12 shows the photocurrent values of Ti/TiO$_2$NTEG-3h-500-Pt-Xcycles by the choronoamperometric method. As the amount of loaded Pt amount on the nanotube surface increases, the photocurrent values decrease gradually. This could be due to the absorbed photon amount on TiO$_2$ surface which decreases by increasing Pt amount.

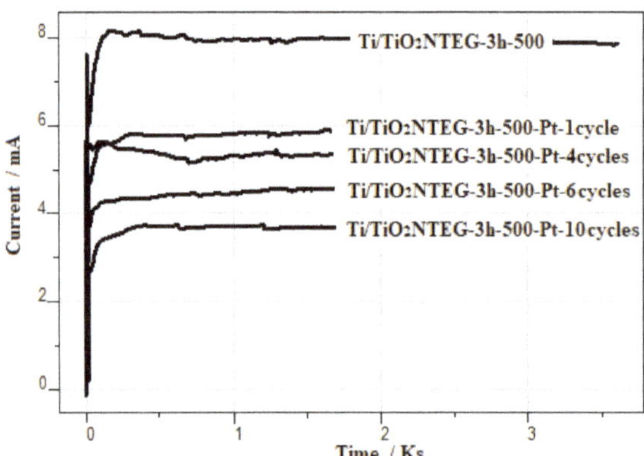

Figure 12. The photocurrent values of Ti/TiO$_2$NTEG-3h-500 and Ti/TiO$_2$NTEG-3h-500-Pt-Xcycles electrodes in the same medium in which PEC paraquat degradation reactions were performed.

3.2. Photoelectrocatalytic Activity Results

Preliminary experiments showed that 1.0 V was the optimum bias for the PEC degradation of paraquat (see Figure S13), thus all PEC and EC experiments were performed at 1.0 V.

Table 3 shows the results of PEC degradation of paraquat using Ti/TiO$_2$-500 and Ti/TiO$_2$NTHF-1h-Y photoanodes. Except for Ti/TiO$_2$NTHF-1h-750, all Ti/TiO$_2$NTHF-1h-Y photoanodes showed higher PEC activity than Ti/TiO$_2$-500, probably because the nanotube structured electrodes have higher surface area. Ti/TiO$_2$NTHF-1h-750 did not show any PEC activity, due to the very high calcination temperature, which probably caused the loss of hydroxyl groups responsible of oxygen adsorption [9]. Oxygen is necessary, as it is an electron acceptor for the initial step of the redox reaction [36]. Ti/TiO$_2$NT-1h-500 showed the highest paraquat conversion for 3 h PEC reaction time. Indeed, its conversion is almost two-fold higher than Ti/TiO$_2$-500 (30% vs. 17%).

Table 3. PEC degradation results of paraquat by using Ti/TiO$_2$-500 and Ti/TiO$_2$NTHF-1h-Y electrodes. [paraquat]: 37.4 μM, Applied potential vs. Ag/AgCl: 1.0 V.

Photoanode	Calcination Temperature (°C)	Conversion for 3 h (%)
Ti/TiO$_2$-500	500	17
Ti/TiO$_2$NTHF-1h-400	400	28
Ti/TiO$_2$NTHF-1h-500	500	30
Ti/TiO$_2$NTHF-1h-600	600	23
Ti/TiO$_2$NTHF-1h-650	650	26
Ti/TiO$_2$NTHF-1h-750	750	0

Table 4 shows PEC paraquat degradation results of Ti/TiO$_2$NTEG-Xh-500, Ti/TiO$_2$NTHF-1h-500 and Ti/TiO$_2$NTHF-6h-500 electrodes. Results of representative PEC experiments are shown in Figure S14. Ti/TiO$_2$NTEG-1h-500 exhibited almost three times higher conversion (ca. 86% versus 30%) than Ti/TiO$_2$NTHF-1h-500 for 3h PEC reaction time. This result is due to the formation of longer nanotube structures in the EG medium, instead of in HF, as SEM images suggest.

Nanotube pores are located in vertical position with respect to the irradiation. Therefore, the light could reach the inside of the pores. For this reason, anodes with nanotube structured TiO$_2$ showed much more activity compared to non-nanotube structured one (Ti/TiO$_2$-500). In addition, the anodes with long nanotubes showed higher PEC activity than shorter ones. However, due to the diffraction and reflection of photons inside of the tube, the intensity of light decreases as it travels through the tube [37].

The effect of the anodic oxidation time on the preparation of Ti/TiO$_2$NTEG-Xh-500 electrodes was investigated and the best activity was achieved by Ti/TiO$_2$NTEG-3h-500 photoanode. Figure S15 shows UV-Vis absorbance values of the samples taken from the reaction medium at fixed times during PEC degradation of paraquat by using the Ti/TiO$_2$NTEG-3h-500 photoanode. As it can be noticed by the observation of the SEM results (see Table 2), the length of the TiO$_2$ nanotubes increased considerably by increasing the anodic oxidation time. A drawback can be the fact that, for very long nanotubes, the light cannot irradiate the whole internal surface. Indeed, the results of Ti/TiO$_2$NTEG-4h-500 and Ti/TiO$_2$NTEG-3h-500 are similar. In addition, the Ti/TiO$_2$NTEG-6h-500 electrode showed a lower activity, probably because parts of the nanotubes are covered by a layer of TiO$_2$, thus causing a decrease of the active surface area as shown in the SEM image (Figure 8d). Furthermore, since its nanotube length is very high (11 μm), the mass transfer of paraquat at the bottom of the tube could be limited.

Table 4. The results of PEC, PC and EC paraquat degradation by using Ti/TiO$_2$NTEG-Xh-500 photoelectrodes. [paraquat]: 37.4 μM, applied potential vs. Ag/AgCl: 1.0 V.

Photoanode	Method	Anodic Oxidation Time	Conversion for 1 h (%)	Conversion for 3 h (%)
Ti/TiO$_2$NTEG-1h-500	PEC	1	43	86
Ti/TiO$_2$NTEG-2h-500	PEC	2	52	93
Ti/TiO$_2$NTEG-3h-500	PEC	3	60	98
Ti/TiO$_2$NTEG-4h-500	PEC	4	62	95
Ti/TiO$_2$NTEG-6h-500	PEC	6	50	90
Ti/TiO$_2$NTEG-6h-500	PC	6	7	26
Ti/TiO$_2$NTEG-6h-500	EC	6	0	1
Ti/TiO$_2$NTHF-1h-500	PEC	1	8	30
Ti/TiO$_2$NTHF-6h-500	PEC	6	9	48

The comparison between electrodes prepared in different solvents (e.g., EG or HF solution) shows that, by using EG, a higher activity was obtained because longer nanotubes formed.

Moreover, it can be noticed that runs carried out by using only PC and EC showed very low activities (Table 4 and Figure 13). Therefore, it is evident that photoelectrocatalysis shows a high synergy between PC and EC methods, by reducing the recombination rate of the electron–hole pairs formed by UV irradiation. In other words, by decreasing the recombination rate, the possibility of the interaction of electron–hole pairs with suitable species increases.

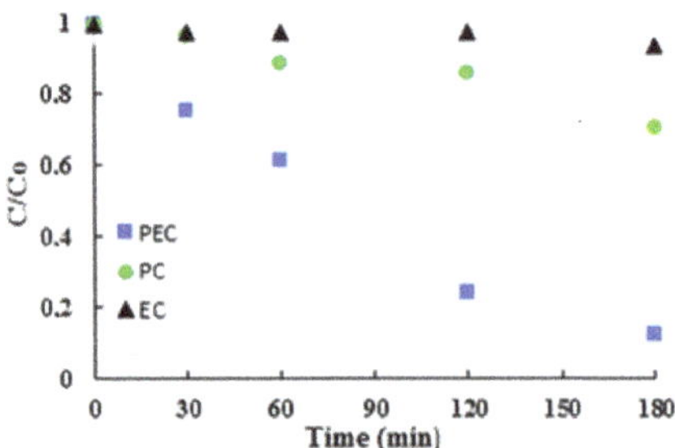

Figure 13. Experimental results of PC, EC and PEC degradation of paraquat by using Ti/TiO$_2$NTEG-6h-500 photoanode. [paraquat]: 37.4 µM, applied potential vs. Ag/AgCl: 1.0 V.

Table 5 reports the results of PEC paraquat degradation experiments carried out by using Ti/TiO$_2$NTEG-3h-500-Pt-Xcycles photoanodes. By increasing the number of cycles used for Pt loading, the Pt amount increased together with the size of Pt particles, as shown in SEM images. Ti/TiO$_2$NTEG-3h-500-4cycles photoanode showed the best performance (75% conversion), probably due to an optimal Pt dispersion. With bigger Pt nanoparticle size, the activity decreased probably due to a reduction of the effective surface area of the photoanode. Non-platinized one (Ti/TiO$_2$NTEG-3h-500) showed 60% conversion for 1 h reaction time.

Table 5. PEC experiment results of Ti/TiO$_2$NTEG-3h-500-Pt-Xcycles photoanodes. [paraquat]: 37.4 µM, applied potential vs. Ag/AgCl: 1.0 V.

Photoanode	Cycle Count	Conversion for 1 h (%)
Ti/TiO$_2$NTEG-3h-500	-	60
Ti/TiO$_2$NTEG-3h-500-Pt-1cycle	1	49
Ti/TiO$_2$NTEG-3h-500-Pt-3cycles	3	60
Ti/TiO$_2$NTEG-3h-500-Pt-4cycles	4	75
Ti/TiO$_2$NTEG-3h-500-Pt-5cycles	5	68
Ti/TiO$_2$NTEG-3h-500-Pt-6cycles	6	60
Ti/TiO$_2$NTEG-3h-500-Pt-7cycles	7	52
Ti/TiO$_2$NTEG-3h-500-Pt-10cycles	10	46
Ti/TiO$_2$NTHF-3h-500-Pt-25cycles	25	42

4. Conclusions

Effective nanotube structured TiO$_2$ on Ti plate photoanodes in ethylene glycol medium (Ti/TiO$_2$-NTEG) was prepared, characterized and used for photoelectrocatalytic degradation of paraquat, which is one of the most used herbicides. Thermally oxidized TiO$_2$ on Ti plate (Ti/TiO$_2$-500) and nanotube structured TiO$_2$ on Ti plates prepared in HF medium (Ti/TiO$_2$-NTHF) were also prepared and used for comparison. Ti/TiO$_2$-NTEG photoanodes were also loaded by Pt nanoparticles by cyclic voltammetry method. The effects of nanotube length and Pt amount of photoanodes on the activity were investigated. The obtained results show that Ti/TiO$_2$-NTEG photoanodes consists of very long TiO$_2$ nanotubes which raise the activity, therefore they showed much higher PEC activity than other type of used electrodes.

According to XRD analysis, all Ti/TiO$_2$-NTEG-Xh-Y electrodes are in anatase phase with negligible amount of rutile whose amount increased by increasing thermal treatment temperature. SEM and XRD results show that loaded Pt in photoanode is in the metallic form and in cubic structure. In addition, Pt nanoparticles grow with increasing number of Pt loading cycles.

The primary particle sizes of the anatase peak of all Ti/TiO$_2$-NTEG electrodes are close to each other and are about 35 nm. Interestingly, this value is approximately the same the wall thickness of the TiO$_2$ nanotubes of Ti/TiO$_2$-NTEG photoanodes. We also found a linear correlation between nanotube length and XRD peak intensity.

Photocurrent values of Ti/TiO$_2$NTEG photoanode are higher than that of Ti/TiO$_2$NTHF. Moreover, by increasing loaded Pt amount on Ti/TiO$_2$NTEG photoanode, photocurrent values decrease linearly due to the reduced absorbed photon amount on the TiO$_2$ surface.

Ti/TiO$_2$-NTEG-6h-500 has the longest nanotube, however a portion of the nanotube surface is covered by TiO$_2$ layers. These layers had a detrimental effect on the photoelectrocatalytic activity because they reduced the effective surface area of the material.

Ti/TiO$_2$-NTEG-3h-500-Pt-4cycles photoanode showed highest PEC activity for paraquat degradation. We obtained a significant synergy for PEC reaction of paraquat as the PC oxidation reaction was slow and especially almost no EC activity was obtained. Indeed, PEC process reduced the recombination rate of the electron–hole pairs formed by UV irradiation on the photoanode surface.

Supplementary Materials: The following are available online at http://www.mdpi.com/1996-1944/11/9/1715/s1, Figure S1: The photo of Ti/TiO$_2$-500 photoanode, Figure S2: Experimental setup used for anodic oxidation, Figure S3: The photos of Ti/TiO$_2$NTHF-X-Y photoanodes, Figure S4: The photos of Ti/TiO$_2$NTEG-X-Y photoanodes, Figure S5: Voltammograms obtained during Pt nanoparticle loading on Ti/TiO$_2$NTEG-3h-500 electrode by CV until 4 cycles. The first cycle is black, the second is green, the third is red and the fourth is blue, Figure S6: SEM images of Ti/TiO$_2$NTHF-6h-650 photoanode (magnification: 1000× (a), and 5000× (b)), Figure S7: PEC experiment system (up) and the spectra of the used UV fluorescent lamp (below), Figure S8: XRD patterns of Ti/TiO$_2$NTEG500-3h-Pt-25cycles electrode, Figure S9: SEM image of Ti/TiO$_2$-500 photoanode. Magnification: 50,000× Figure S10: SEM image of Ti/TiO$_2$NTHF-1h-500 photoanode (magnification: 250,000×), Figure S11: SEM images of Ti/TiO$_2$NTEG-4h-500 photoanode. Cross section view (magnification: 10,000×), Figure S12: SEM images of Ti/TiO$_2$NTEG-6h-500 photoanode. Cross section view (magnification: 150,000×). Bottom view (magnification: 20,000×), Figure S13: The conversion values for PEC paraquat degradation for 1 (blue) and 3 (green) hours of reaction time at different voltage values in the presence of Ti/TiO$_2$NTEG-3h-500 electrode, Figure S14. The PEC experiment results of Ti/TiO$_2$-500 (▬), Ti/TiO$_2$NTHF-1h-500 (●), Ti/TiO$_2$NTEG-1h-500 (▲), Ti/TiO$_2$NTEG-3h-500 (■) and Ti/TiO$_2$NTEG-6h-500 (◆) for paraquat degradation. Potential: 1V, Figure S15. UV-Vis absorbance values of the samples taken from the reaction medium at fixed times during PEC degradation of paraquat (37.4 μM) at 1 V by using the Ti/TiO$_2$NTEG-3h-500 photoanode.

Author Contributions: L.Ö. and S.Y. designed the experiments; T.M. prepared the photoanodes and performed the PEC, PC and EC experiments; and S.Y. and L.Ö. contributed to analyzing the data and writing the paper.

Funding: This research was funded by the Scientific Research Project Council of Afyon Kocatepe University (AKÜ-BAP), Turkey, with grant number [16.KARİYER.172].

Acknowledgments: The authors thank Vittorio Loddo (Università degli Studi di Palermo, Italy) for useful suggestions, Erhan Karaca (İLTEM, Dumlupınar Üniversitesi, Turkey) for SEM-EDX analyses, Yasemin Çimen (Eskişehir Teknik Üniversitesi, Turkey) for SEM analyses (Figures 4–6 and S6) and Hakan Şahin (TUAM, Afyon Kocatepe Üniversitesi, Turkey) for XRD analyses.

Conflicts of Interest: The authors declare no conflicts of interest.

References

1. Marien, C.B.D.; Cottineau, T.; Robert, D.; Drogui, P. TiO$_2$ Nanotube arrays: Influence of tube length on the photocatalytic degradation of Paraquat. *Appl. Catal. B Environ.* **2016**, *194*, 1–6. [CrossRef]

2. Kowal, S.; Balsaa, P.; Werres, F.; Schmidt, T.C. Determination of the polar pesticide degradation product N, N-dimethylsulfamide in aqueous matrices by UPLC–MS/MS. *Anal. Bioanal. Chem.* **2009**, *395*, 1787–1794. [CrossRef] [PubMed]

3. Zahedi, F.; Behpour, M.; Ghoreishi, S.M.; Khalilian, H. Photocatalytic degradation of paraquat herbicide in the presence TiO$_2$ nanostructure thin films under visible and sun light irradiation using continuous flow photoreactor. *Sol. Energy* **2015**, *120*, 287–295. [CrossRef]

4. Daghrir, R.; Drogui, P.; Robert, D. Photoelectrocatalytic technologies for environmental applications. *J. Photochem. Photobiol. A Chem.* **2012**, *238*, 41–52. [CrossRef]

5. Suhadolnik, L.; Pohar, A.; Likozar, B.; Ceh, M. Mechanism and kinetics of phenol photocatalytic, electrocatalytic and photoelectrocatalytic degradation in a TiO_2-nanotube fixed-bed microreactor. *Chem. Eng. J.* **2016**, *303*, 292–301. [CrossRef]

6. Serpone, N.; Pelizzetti, E. *Photocatalysis: Fundamentals and Applications*; Wiley: New York, NY, USA, 1989; ISBN 9780471967545-20160527.

7. Fujishima, A.; Zhang, X.; Tryk, D.A. TiO_2 photocatalysis and related surface phenomena. *Surf. Sci. Rep.* **2008**, *63*, 515–582. [CrossRef]

8. Augugliaro, V.; Camera-Roda, G.; Loddo, V.; Palmisano, G.; Palmisano, L.; Soria, J.; Yurdakal, S. Heterogeneous photocatalysis and photoelectrocatalysis: From unselective abatement of noxious species to selective production of high-value chemicals. *J. Phys. Chem. Lett.* **2015**, *6*, 1968–1981. [CrossRef] [PubMed]

9. Özcan, L.; Yurdakal, S.; Augugliaro, V.; Loddo, V.; Palmas, S.; Palmisano, G.; Palmisano, L. Photoelectrocatalytic selective oxidation of 4-methoxybenzyl alcohol in water by TiO_2 supported on titanium anodes. *Appl. Catal. B Environ.* **2013**, *132*, 535–542. [CrossRef]

10. Rajeshwar, K. Photoelectrochemistry and the environment. *J. Appl. Electrochem.* **1995**, *25*, 1067–1082. [CrossRef]

11. Palmisano, G.; Loddo, V.; El Nazer, H.H.; Yurdakal, S.; Augugliaro, V.; Ciriminna, R.; Pagliaro, M. Graphite-supported TiO_2 for 4-nitrophenol degradation in a photoelectrocatalytic reactor. *Chem. Eng. J.* **2009**, *155*, 339–346. [CrossRef]

12. Liu, D.; Yang, T.; Chen, J.; Chou, K.C.; Hou, X. Pt-Co alloys-loaded cubic SiC electrode with improved photoelectrocatalysis property. *Materials* **2017**, *10*, 955. [CrossRef] [PubMed]

13. Ensaldo-Rentería, M.K.; Ramírez-Robledo, G.; Sandoval-González, A.; Pineda-Arellano, C.A.; Álvarez-Gallegos, A.A.; Zamudio-Lara, Á.; Silva-Martínez, S. Photoelectrocatalytic oxidation of acid green 50 dye in aqueous solution using Ti/TiO_2-NT electrode. *J. Environ. Chem. Eng.* **2018**, *6*, 1182–1188. [CrossRef]

14. Chai, S.; Zhao, G.; Li, P.; Lei, Y.; Zhang, Y.N.; Li, D. Novel sieve-like SnO_2/TiO_2 nanotubes with integrated photoelectrocatalysis: Fabrication and application for efficient toxicity elimination of nitrophenol wastewater. *J. Phys. Chem. C* **2011**, *115*, 18261–18269. [CrossRef]

15. Zhu, H.; Zhao, M.; Zhou, J.; Li, W.; Wang, H.; Xu, Z.; Lu, L.; Pei, L.; Shi, Z.; Yan, S.; et al. Surface states as electron transfer pathway enhanced charge separation in TiO_2 nanotube water splitting photoanodes. *Appl. Catal. B Environ.* **2018**, *234*, 100–108. [CrossRef]

16. Mohapatra, S.K.; Misra, M.; Mahajan, V.K.; Raja, K.S. A novel method for the synthesis of titania nanotubes using sonoelectrochemical method and its application for photoelectrochemical splitting of water. *J. Catal.* **2007**, *246*, 362–369. [CrossRef]

17. Bettoni, M.; Rol, C.; Sebastiani, G.V. Photoelectrochemistry on TiO_2/Ti anodes as a tool to increase the knowledge about some photo-oxidation mechanisms in CH_3CN. *J. Phys. Org. Chem.* **2008**, *21*, 219–224. [CrossRef]

18. Özcan, L.; Yalçın, P.; Alagöz, O.; Yurdakal, S. Selective photoelectrocatalytic oxidation of 5-(hydroxymethyl)-2-furaldehyde in water by using Pt loaded nanotube structure of TiO_2 on Ti photoanodes. *Catal. Today* **2017**, *281*, 205–213. [CrossRef]

19. Jia, Y.; Ye, L.; Kang, X.; You, H.; Wang, S.; Yao, J. Photoelectrocatalytic reduction of perchlorate in aqueous solutions over Ag doped TiO_2 nanotube arrays. *J. Photochem. Photobiol. A Chem.* **2016**, *328*, 225–232. [CrossRef]

20. Assefpour-Dezfuly, M.; Vlachos, C.; Andrews, E.H. Oxide morphology and adhesive bonding on titanium surfaces. *J. Mater. Sci.* **1984**, *19*, 3626–3639. [CrossRef]

21. Zwilling, V.; Darque-Ceretti, E.; Boutry-Forveille, A.; David, D.; Perrin, M.Y.; Aucouturier, M. Structure and physicochemistry of anodic oxide films on titanium and TA6V alloy. *Surf. Interface Anal.* **1999**, *27*, 629–637. [CrossRef]

22. Gong, D.; Grimes, C.A.; Varghese, O.K.; Hu, W.C.; Singh, R.S.; Chen, Z.; Dickey, E.C. Titanium oxide nanotube arrays prepared by anodic oxidation. *J. Mater. Res.* **2001**, *16*, 3331–3334. [CrossRef]

23. Macak, J.M.; Sirotna, K.; Schmuki, P. Self-organized porous titanium oxide prepared in Na_2SO_4/NaF electrolytes. *Electrochim. Acta* **2005**, *50*, 3679–3684. [CrossRef]

24. Yasuda, K.; Schmuki, P. Control of morphology and composition of self-organized zirconium titanate nanotubes formed in $(NH_4)_2SO_4$/NH_4F electrolytes. *Electrochim. Acta* **2007**, *52*, 4053–4061. [CrossRef]

25. Paulose, M.; Prakasam, H.E.; Varghese, O.K.; Peng, L.; Popat, K.C.; Mor, G.K.; Desai, T.A.; Grimes, C.A. TiO_2 nanotube arrays of 1000 μm length by anodization of titanium foil: Phenol red diffusion. *J. Phys. Chem. C* **2007**, *111*, 14992–14997. [CrossRef]

26. Yoriya, S.; Grimes, C.A. Self-assembled TiO_2 nanotube arrays by anodization of titanium in diethylene glycol: Approach to extended pore widening. *Langmuir* **2010**, *26*, 417–420. [CrossRef] [PubMed]

27. Truong, N.N.; Altomare, M.; Yoo, J.; Schmuki, P. Efficient photocatalytic H_2 evolution: Controlled dewetting-dealloying to fabricate site-selective high-activity nanoporous Au particles on highly ordered TiO_2 nanotube arrays. *Adv. Mater.* **2015**, *27*, 3208–3215. [CrossRef]

28. Xie, K.P.; Sun, L.; Wang, C.L.; Lai, Y.K.; Wang, M.Y.; Chen, H.B.; Lin, C.J. Photoelectrocatalytic properties of Ag nanoparticles loaded TiO_2 nanotube arrays prepared by pulse current deposition. *Electrochim. Acta* **2010**, *55*, 7211–7218. [CrossRef]

29. Qin, Y.H.; Yang, H.H.; Lv, R.L.; Wang, W.G.; Wang, C.W. TiO_2 nanotube arrays supported Pd nanoparticles for ethanol electrooxidation in alkaline media. *Electrochim. Acta* **2013**, *106*, 372–377. [CrossRef]

30. Zhang, L.; Pan, N.Q.; Lin, S.W. Influence of Pt deposition on water-splitting hydrogen generation by highly-ordered TiO_2 nanotube arrays. *Int. J. Hydrogen Energy* **2014**, *39*, 13474–13480. [CrossRef]

31. Ge, M.Z.; Cao, C.Y.; Huang, J.Y.; Li, S.H.; Zhang, S.N.; Deng, S.; Li, O.S.; Zhang, K.Q.; Lai, Y.K. Synthesis, modification, and photo/photoelectro catalytic degradation applications of TiO_2 nanotube arrays: A review. *Nanotechnol. Rev.* **2016**, *5*, 75–112. [CrossRef]

32. Moctezuma, E.; Leyva, E.; Monreal, E.; Villegas, N.; Infante, D. Photocatalytic degradation of the herbicide "paraquat". *Chemosphere* **1999**, *39*, 511–517. [CrossRef]

33. Kanchanatip, E.; Grisdanurak, N.; Thongruang, R.; Neramittagapong, A. Degradation of paraquat under visible light over fullerene modified V-TiO_2. *React. Kinet. Mech. Cat.* **2011**, *103*, 227–237. [CrossRef]

34. Shankar, K.; Mor, G.K.; Prakasam, H.E.; Yoriya, S.; Paulose, M.; Varghese, O.K.; Grimes, C.A. Highly-ordered TiO_2 nanotube arrays up to 220 μm in length: Use in water photoelectrolysis and dye-sensitized solar cells. *Nanotechnology* **2007**, *18*, 065707. [CrossRef]

35. Yurdakal, S.; Tek, B.S.; Alagöz, O.; Augugliaro, V.; Loddo, V.; Palmisano, G.; Palmisano, L. Photocatalytic selective oxidation of 5-(hydroxymethyl)-2-furaldehyde to 2,5-furandicarbaldehyde in water by using anatase, rutile, and brookite TiO_2 nanoparticles. *ACS Sustain. Chem. Eng.* **2013**, *1*, 456–461. [CrossRef]

36. Schiavello, M. *Heterogeneous Photocatalysis*; Wiley: Chichester, UK, 1997; ISBN 0-471-96754-8.

37. Loddo, V.; Yurdakal, S.; Palmisano, G.; Imoberdorf, G.E.; Irazoqui, H.A.; Alfano, O.M.; Augugliaro, V.; Berber, H.; Palmisano, L. Selective photocatalytic oxidation of 4-methoxybenzyl alcohol to p-anisaldehyde in organic-free water in a continuous annular fixed bed reactor. *Int. J. Chem. React. Eng.* **2007**, *5*, A57. [CrossRef]

 *materials* 

Article

Enhanced Photocatalytic and Antimicrobial Performance of Cuprous Oxide/Titania: The Effect of Titania Matrix

Marcin Janczarek [1,2,3,*], Maya Endo [1], Dong Zhang [1], Kunlei Wang [1] and Ewa Kowalska [1]

[1] Institute for Catalysis, Hokkaido University, Sapporo 001-0021, Japan; m_endo@cat.hokudai.ac.jp (M.E.); zhang.d@cat.hokudai.ac.jp (D.Z.); kunlei@cat.hokudai.ac.jp (K.W.); kowalska@cat.hokudai.ac.jp (E.K.)
[2] Department of Chemical Technology, Gdansk University of Technology, 80-233 Gdansk, Poland
[3] Institute of Chemical Technology and Engineering, Faculty of Chemical Technology, Poznan University of Technology, 60-965 Poznan, Poland
* Correspondence: marcin.janczarek@put.poznan.pl

Received: 12 September 2018; Accepted: 19 October 2018; Published: 23 October 2018

Abstract: A simple, low-cost method was applied to prepare hybrid photocatalysts of copper (I) oxide/titania. Five different TiO_2 powders were used to perform the study of the effect of titania matrix on the photocatalytic and antimicrobial properties of prepared nanocomposites. The photocatalytic efficiency of such a dual heterojunction system was tested in three reaction systems: ultraviolet-visible (UV-Vis)-induced methanol dehydrogenation and oxidation of acetic acid, and 2-propanol oxidation under visible light irradiation. In all the reaction systems considered, the crucial enhancement of photocatalytic activity in relation to corresponding bare titania was observed. The reaction mechanism for a specific reaction and the influence of titania matrix were discussed. Furthermore, antimicrobial (bactericidal and fungicidal) properties of Cu_2O/TiO_2 materials were analyzed. The antimicrobial activity was found under UV, visible and solar irradiation, and even for dark conditions. The origin of antimicrobial properties with emphasis on the role of titania matrix was also discussed.

Keywords: photocatalysis; nanocomposites; heterojunction; Z-scheme; Cu_2O; TiO_2; antimicrobial properties

---

1. Introduction

Titanium dioxide (TiO_2, titania) is widely recognized as an efficient, stable and green photocatalytic material (long-term stability, chemical inertness, corrosion resistance and non-toxicity). Therefore, its application potential in photocatalysis is still growing and presently focused on areas such as environmental remediation (water treatment and air purification), renewable energy processes (i.e., water splitting for hydrogen production, conversion of CO_2 to hydrocarbons), and self-cleaning surfaces [1–3]. However, one can distinguish two main problems in the wider application of TiO_2. Firstly, the application of titania is still limited to the regions with a high intensity of solar radiation due to its wide bandgap (ca. 3.0 to 3.2 eV). Strategies such as titania doping [4], surface modification [5], semiconductor coupling [6], and dye sensitization [7] can be applied to incorporate visible light absorption to TiO_2. Another important limitation decreasing photocatalytic activity is the recombination of the photogenerated electron-hole pairs caused by impurities, defects and other factors, which introduce bulk or surface imperfections into the titania crystal. The solution is the incorporation of species capable of promoting charge separation (e.g., TiO_2 modification with metal ions, noble metals and heterojunction coupling with other semiconductors). Taking into consideration both described limitations, the methods to improve photocatalytic activity of TiO_2 are similar. Therefore, a proper method selection to rectify both limitations is important to prepare a photocatalytic material with universal properties, active both in ultraviolet (UV) and visible light [8,9].

Copper (Cu) as a candidate for titania modification is a very promising material [10,11]. The first reason is the advantageous price and well-known antimicrobial properties. In comparison to other noble metals (gold, platinum and silver), recognized as very efficient co-catalysts of titania, copper, as the consequence of its abundance in the Earth's crust, is an inexpensive material, 100 times and 6000 times cheaper than silver and gold, respectively. Copper can exist in the following oxidation states, i.e., Cu^0, Cu^I and Cu^{II}, and therefore the active copper species in TiO_2 photocatalytic system can be recognized as copper oxides (Cu_2O, CuO) and metallic copper. Copper oxidation states can also change as the consequence of sample drying [12–14] and reaction conditions [11,15]. For example, despite zero-valent copper being easily formed on the titania surface either by photodeposition [12–14] or radiolytic reduction [16], the contact with air results in fast oxidation of the copper, and the resultant photocatalysts possess different forms of co-existing copper species (mainly metallic core and oxide shell). It should be pointed that although copper is easily oxidized, stable zero-valent copper has been also reported when stabilized by titania aerogel [17]. Considering the above issues (the variety of copper forms and their relative instability), there is a difficulty in understanding their role in different reaction systems. Among copper oxides in relation to heterojunction with titania, Cu_2O is one of the few p-type semiconductors which are inexpensive, non-toxic and widely available. The Cu_2O/TiO_2 p-n heterojunction system has promising application potential both in the oxidation of organic pollutants including very good antipathogenic properties [15,18–32] and in photocatalytic hydrogen production [11,29,33–39].

For example, Bessekhouad et al. proposed that under visible light irradiation, electrons from Cu_2O were injected into the conduction band (CB) of TiO_2, and at the titania surface could react with dissolved oxygen molecules inducing the formation of oxygen peroxide radicals ($O_2{}^{\cdot-}$) [21]. In the case of a UV system, an increase in the content of Cu_2O resulted in enhanced efficiency, but resultant activities at high content of Cu_2O were only slightly higher than that of pure TiO_2 [21]. Similarly, Huang et al. found the improvement of photocatalytic activity for a Cu_2O/TiO_2 system induced by UV and visible light, i.e., 6 and 27 times higher photocatalytic activity than that for pure P25, respectively [30,31]. Moreover, it was found that an increase in the content of Cu_2O resulted in higher photocatalytic activity (the highest activity for 70%-Cu_2O content). A Cu_2O/TiO_2 p-n heterojunction system was also successfully applied for photocatalytic hydrogen generation. Zhang et al. prepared Cu_2O/TiO_2 composites through the deposition of copper on titania nanotube arrays [29]. Since the CB of Cu_2O is more negative than that of TiO_2, the excited electrons are quickly transferred from Cu_2O nanoparticles to titania, leaving the holes on the valence band (VB) of Cu_2O and leading to an effective reduction of protons to H2.

Moreover, copper, especially copper (I) oxide, has been well known as a antimicrobial agent since ancient times. Due to its advantages, e.g., inexpensiveness, low toxicity and abundant sources, it has been applied to improve the photo-induced antimicrobial activity of titania. The proposed mechanisms include: (i) the structure of surface proteins are denatured [40]; and (ii) the adsorbed copper ions induce oxidative stress in the bactericidal process [41], and the accumulation of copper ions inside bacteria [42]. It was found that the optimal balance of Cu_2O and CuO content in Cu_xO/TiO_2 composite photocatalyst was important to achieve good antibacterial performance under visible light irradiation and dark conditions and, furthermore, Cu_2O/TiO_2 was reported to be more active than CuO/TiO_2 and $CuNPs/TiO_2$ [18].

To the authors' best knowledge there is still no comprehensive research paper considering the effect of a titania matrix in Cu_2O-titania heterojunction system for photocatalytic and antimicrobial properties. The clarification of this issue is important in order to analyze application perspectives of such photocatalysts. In this study, to evaluate the role of titania matrix, five types of titania were considered. Heterojunctions between Cu_2O and different TiO_2 were prepared as physical mixtures of powders. Enhanced photocatalytic properties were discussed based on three reaction systems:

methanol dehydrogenation and oxidation of acetic acid under UV/vis irradiation, and 2-propanol oxidation under vis irradiation. The antimicrobial properties of prepared composite photocatalysts were tested by using *E. coli* and *C. albicans*.

2. Materials and Methods

2.1. Preparation of Cu_2O/TiO_2 Photocatalysts

TiO_2 samples, selected for the preparation procedure, were supplied by several sources: P25 (AEROXIDE® TiO_2 P25, Nippon Aerosil, Yokkaichi, Japan), ST-01 (ST-01, Ishihara Sangyo, Yokkaichi, Japan) ST-41 (ST-41, Ishihara Sangyo, Yokkaichi, Japan), TIO-6 (TIO-6, Catalysis Society of Japan, Tokyo, Japan), RUT (rutile nanopowder, Sigma-Aldrich, Saint Louis, MO, USA). Cu_2O was supplied by Wako Pure Chemicals, Tokyo, Japan. All materials were used as received, without further processing. Cu_2O/TiO_2 composites were prepared by physical mixing of Cu_2O and TiO_2 powders in an agate mortar. Titania samples were mixed with different contents of Cu_2O resulting in preparation of the composites containing 1, 5, 10 and 50 wt % of Cu_2O. The time of grinding (5 min) was the same for all samples to ensure appropriate homogeneity of prepared composite powders.

2.2. Characterization

The ultraviolet-visible (UV-Vis) diffuse reflectance spectra (DRS, JASCO, Tokyo, Japan) were recorded on JASCO V-670 equipped with PIN-757 integrating sphere using $BaSO_4$ as a reference. Gas-adsorption measurements of prepared titania samples were performed on a Yuasa Ionics Autosorb 6AG surface area and pore size analyzer, Osaka, Japan. Specific surface area (SSA) was calculated from nitrogen adsorption at 77 K using the Brunauer–Emmett–Teller equation. X-ray diffraction (XRD) patterns were collected using an X-ray diffractometer (Rigaku intelligent XRD SmartLab with a Cu target, Tokyo, Japan).

2.3. Photocatalytic Activity Tests

The photocatalytic activity of prepared photocatalysts was tested in three reaction systems: (1) decomposition of acetic acid under UV/vis irradiation, (2) dehydrogenation of methanol under UV/vis irradiation, and (3) oxidation of 2-propanol under vis irradiation ($\lambda > 420$ nm: Xe lamp, water IR filter, cold mirror and cut-off filter Y45). For activity testing, 50 mg of photocatalyst was suspended in 5 mL of aqueous solution of (1) methanol (50 vol %), (2) acetic acid (5 vol %), and (3) 2-propanol (5 vol %). The methanol dehydrogenation system was also tested in the presence of platinum (samples: Pt/TiO_2, $Pt/Cu_2O/TiO_2$): hydrogen hexachloroplatinate(IV) ($H_2PtCl_6 \cdot 6H_2O$) was added for adjustment to 2 wt % loading on photocatalyst powders. The suspension for reaction (2) was bubbled with argon before irradiation. The 35-mL testing tubes were sealed with rubber septa, continuously stirred and irradiated in a thermostated water bath. Amounts of liberated (1) carbon dioxide in gas phase, (2) hydrogen in gas phase, and (3) acetone in liquid phase (after powder separation) were determined by GC-TCD (1-2) (Shimadzu GC-8A equipped with a thermal conductivity detector, Shimadzu Corp., Kyoto, Japan) and GC-FID (3) (Shimadzu GC-14B equipped with a flame ionization detector, Shimadzu Corp., Kyoto, Japan).

2.4. Antimicrobial Activity Tests with Xenon Lamp Irradiation

Cu_2O/TiO_2 samples, bare titania and cuprous oxide (ca. 7.1 g/L) were dispersed in *Escherichia coli* K12 (ATCC29425) or *Candida albicans* (isolated from patients (throat smear) with immunodeficiency disorders that cause candidiasis (collection from West Pomeranian University of Technology, Szczecin, Poland)) suspension at concentrations of ca. 1–5×10^8 cells/mL (*E. coli* K12) or 1–5×10^4 cells/mL (*C. albicans*) in a test tube with stirring bar, and then irradiated with xenon lamp (with cold mirror (CM2) and UV-D36B filter; $300 < \lambda < 420$ nm or CM1 and Y-45 filter; $\lambda > 420$ nm) or kept in the dark. As a control, bacterial or fungal suspension without titania was also tested. Serial dilutions (10^{-1}–10^{-6}) of microbial suspension were prepared and aliquots of suspensions were inoculated on Plate Count Agar (Becton Dickinson Company, Franklin Lakes, NJ, USA for *E. coli* K12) or Malt Extract Agar (Merck Millipore Corporation, Burlington, MA, USA for *C. albicans*) media at 0, 0.5, 1, 2 and 3 h. Media were incubated at 37 °C overnight, colonies were counted, and the colony-forming unit was determined.

2.5. Antibacterial Activity Test with Solar Irradiation

Cu_2O/TiO_2 (ST-01) or bare titania (ST-01) (0.5 g/L), was dispersed in *E. coli* K12 (ATCC29425) suspension at a concentration of ca. 1×10^8 cells/mL in a glass container with stirring bar, and circulated through glass tubes and irradiated under solar radiation (Sapporo, sunny day, June 2018) or kept in the dark. As a control, bacterial suspension without titania was also tested. The later experimental procedure was the same as that described above.

3. Results and Discussion

3.1. Characterization of Cu_2O/TiO_2 Samples

Five commercially available TiO_2 were selected to perform the study. Table 1 shows the main properties of the samples, which represent diversified types of titania matrix, e.g., different morphology, phase content, crystallite size and specific surface area. Figure 1a shows diffuse reflectance spectra of prepared Cu_2O/TiO_2 with 5 wt %-content of cuprous oxide. The strong absorption at UV range is due to bandgap excitation of titania, and a narrower bandgap of rutile than that of anatase clearly correlates with the absorption edge at longer wavelengths. The absorption peak between 500–600 nm is characteristic of the presence of Cu_2O. Although the content of Cu_2O in each sample was approximately the same, one can observe significant differences in the shape of Cu_2O-absorption region between samples containing a different titania matrix. It is possible to observe the dependency between the size of TiO_2 particles and the height of Cu_2O-originated absorption peak. The Cu_2O/TiO_2 samples with small-particulate titania (ST-01 and TIO-6) are characterized by the strongest 500–600 nm absorption peak of Cu_2O. The crystallite size of Cu_2O determined by XRD analysis was 65 nm and BET surface area: $23 \ m^2 \cdot g^{-1}$. The prevalence of anatase or rutile in a titania matrix did not influence the Cu_2O-originated absorption peak. Another confirmation for the phase presence (anatase, rutile, Cu_2O) in the prepared samples are the XRD results, shown in Figure 2. Cuprous oxide was confirmed in all samples (black patterns after subtraction of titania patterns). Moreover, titania peaks did not change after grinding, which proves that physical mixing was delicate without changing crystal properties (It is known that strong grinding/milling could destroyed titania crystals.). Figure 3 shows scanning transmission electron microscopy (STEM) images with energy dispersive spectroscopy (EDS) mapping, which indicate that Cu_2O particles are uniformly distributed on the surface of titania.

Table 1. TiO_2 samples selected for the study.

Sample Name	Anatase * (%)	Rutile * (%)	Crystallite Size/nm		Specific Surface Area/$m^2 \cdot g^{-1}$
			Anatase	Rutile	
P25	83.8	16.2	21	37	59
ST-01	100	-	8	-	298
ST-41	98.3	1.7	70	124	11
TIO-6	-	100	-	16	105
RUT	1.7	98.3	55	82	4

* Crystalline composition without consideration of amorphous phase.

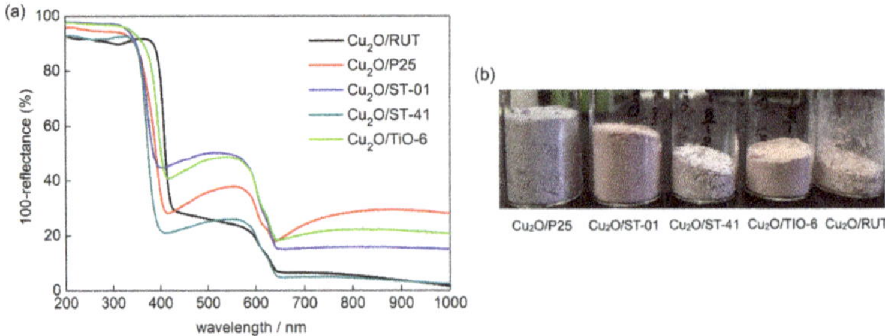

Figure 1. Diffuse reflectance spectra of Cu_2O (5 wt %)/TiO_2 samples (**a**) with respective photographs (**b**).

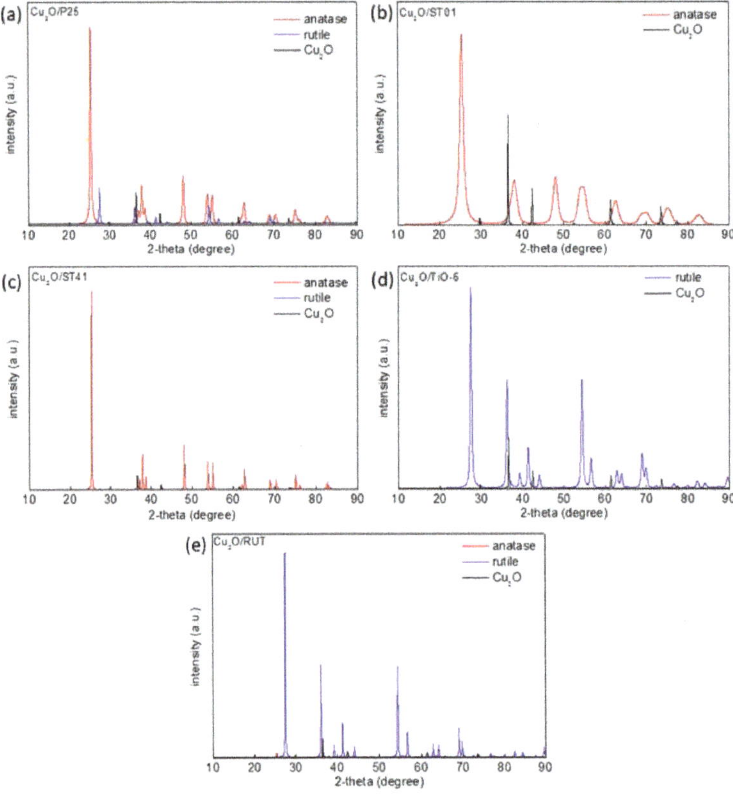

Figure 2. X-ray diffraction (XRD) diffractograms of Cu_2O-modified titania samples: (**a**) Cu_2O/P25, (**b**) Cu_2O/ST-01, (**c**) Cu_2O/ST-41, (**d**) Cu_2O/TIO-6, (**e**) Cu_2O/RUT.

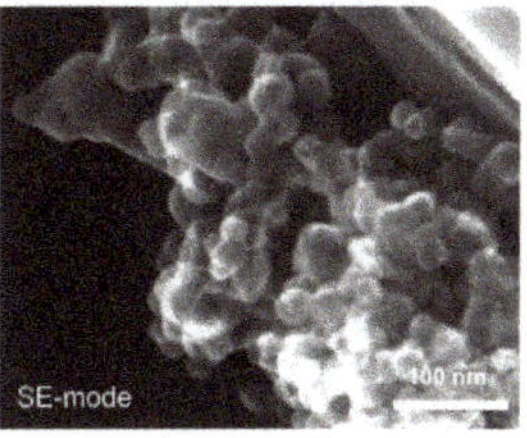

Figure 3. Scanning transmission electron microscopy (STEM) images with energy dispersive spectroscopy (EDS) mapping of Cu_2O/P25 sample. Mapping colors: Ti-violet; Cu-green.

3.2. Photocatalytic Activity

3.2.1. Ultraviolet-Visible (UV-Vis)-Induced Methanol Dehydrogenation

During the reaction of methanol dehydrogenation (H_2 system) under deaerated conditions in the presence of titania and copper oxides the following reactions may be considered:

$$TiO_2 + h\nu \ (\lambda > 290 \ nm) \rightarrow e^- + h^+, \ (generation \ of \ electrons \ and \ holes) \tag{1a}$$

$$Cu_2O + h\nu \ (\lambda > 290 \ nm) \rightarrow e^- + h^+, \tag{1b}$$

$$CH_3OH + h^+ \rightarrow HCHO + H^+, \tag{2}$$

$$Cu^+ + e^- \rightarrow Cu, \tag{3}$$

$$H^+ + e^- \rightarrow 0.5 \ H_2, \tag{4}$$

At the beginning of irradiation, copper oxide is reduced by photoexcited electrons, resulting in the formation of copper deposits on the surface of the photocatalyst (3). In this system, methanol plays the role of a hole scavenger (2), as presented in Figure 4.

As a primary issue, the best copper content (wt %) for Cu_2O/TiO_2 system corresponding to highest photocatalytic activity was determined for further studies. The following copper contents were considered: 1%, 5%, 10% and 50%. For all titania-cuprous oxide heterojunctions, the highest photocatalytic activity was observed for 5 wt % of Cu_2O (Figure 4). By increasing the Cu_2O content, the hydrogen evolution rate decreased because of the increase of charge recombination effect [11] or inner filter effect (competition for photons between two semiconductors). The high content of Cu_2O (50 wt %) caused a significant decrease of photocatalytic activity, probably due to the increase in the opacity and light scattering (shielding effect) influencing photon absorption (irradiation passing through the photocatalyst suspension) [43]. Indeed, characterization of Cu_2O/P25 samples with different content of Cu_2O clearly showed a significant increase in light absorption with an increase in cuprous oxide content at vis range (Figure 5a). Additional XRD analysis confirmed the presence of both titania and cuprous oxide in all hybrid samples (Figure 5b), and the estimated content of cuprous oxide in hybrid materials was almost the same as that used for the preparation of samples (inset in Figure 5b).

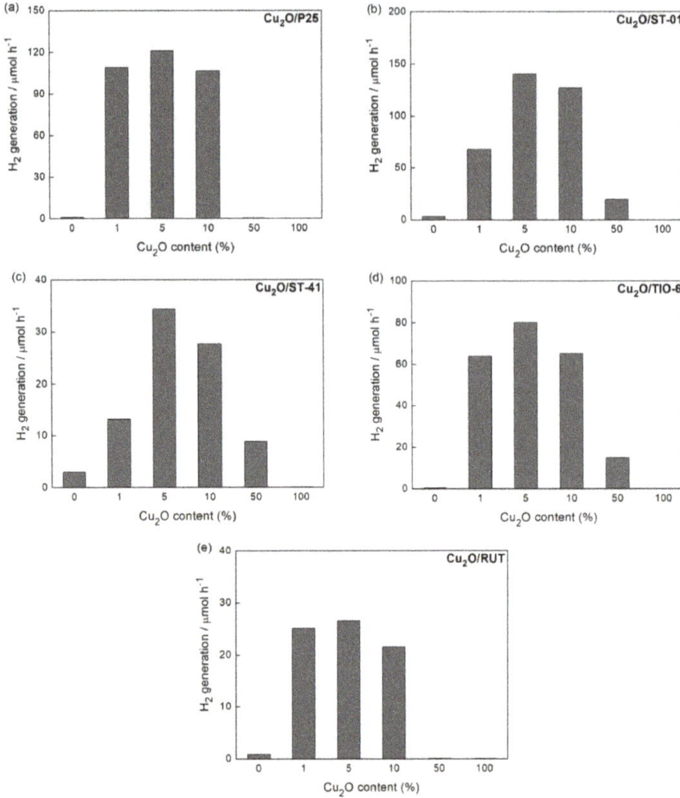

Figure 4. Ultraviolet-visible (UV-Vis) photocatalytic activity of samples: (**a**) Cu$_2$O/P25, (**b**) Cu$_2$O/ST-01, (**c**) Cu$_2$O/ST-41, (**d**) Cu$_2$O/TIO-6, (**e**) Cu$_2$O/RUT, prepared with corresponding Cu$_2$O content in methanol dehydrogenation.

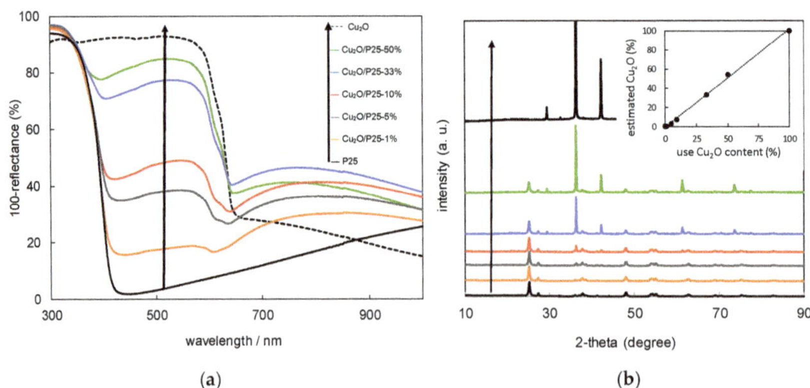

Figure 5. (**a**) Diffuse reflectance spectra and (**b**) XRD diffractograms of Cu$_2$O/P25 samples with different Cu$_2$O content; Inset: Correlation between use and estimated (XRD) content of Cu$_2$O in the samples.

Figure 6 shows the photocatalytic activity in hydrogen system for Cu$_2$O/TiO$_2$ with different titania matrix. These activities were compared with corresponding samples: Pt/TiO$_2$ and Pt/Cu$_2$O/TiO$_2$.

Obviously, the highest photocatalytic activities were obtained for Pt/TiO_2, due to higher work function (6.35 eV vs. 4.7 eV) and smaller overvoltage for hydrogen evolution for platinum than that for copper. Although, the activity of Cu_2O/TiO_2 was 2–3 times lower than that for Pt/TiO_2, it should be remembered that copper is much cheaper than platinum. Moreover, bare titania is practically inactive in this system (Figure 6), and thus evolution of hydrogen on this cheap photocatalyst is quite promising. Additionally, it should be pointed out that another possible mechanism, i.e., type II heterojunction (transfer of photo-generated electrons from CB of Cu_2O to CB of TiO_2 with opposite transfer of photo-generated holes) could be rejected due to the inactivity of bare titania (Figure 6). In contrast to the conclusions of Dozzi et al. [44], the synergistic effect of Pt-Cu was not observed in this reaction system. This is not surprising since Pt-modified titania is one of the most active photocatalysts for hydrogen evolution, and thus formation of other charge carriers' transfers (not only from CB of titania to Pt), e.g., to VB of Cu_2O, should result in hindering of the overall activity. Photocatalytic activities of $Pt/Cu_2O/TiO_2$ were slightly higher than that of Cu_2O/TiO_2 excluding samples based on rutile (TIO-6 and RUT). The higher activity for co-modified anatase samples than that for Cu_2O/TiO_2 could originate from an increase in the content of active sites for hydrogen evolution (both on copper and platinum deposits), whereas the reason for the lowest activity for co-modified rutile samples is unclear. It is possible that more negative CB of rutile than that of anatase (as recently reported [45,46]) could result in two types of co-existing mechanisms (Z-scheme and type II heterojunction), i.e., (1) for Pt-modified titania, photoexcited electrons from CB of Cu_2O migrates to CB of titania (type II heterojunction) and then to Pt deposits (together with directly excited electrons from VB of titania); (2) for Cu_2O/TiO_2 system, Z-scheme mechanism should be preferential (due to inactivity of bare titania); (3) for Pt-modified Cu_2O/TiO_2, similar levels of CBs position for rutile and cuprous oxide could result in the circulation of photogenerated electrons between both semiconductors, instead of their transfer to noble metals' deposits, i.e., $VB(Cu_2O) \rightarrow CB(Cu_2O) \rightarrow CB(rutile) \rightarrow VB(Cu_2O)$. It should be reminded that Pt was deposited in situ, and thus it is highly possible that it could be randomly deposited either on cuprous oxide or on titania. The formation of bimetallic deposits Cu-Pt with metal segregation is also possible, as already reported for titania photocatalysts modified with Au-Cu [47], and Ag-Cu [13,16]. To clarify the mechanism, detailed studies on sample characterization, and reference experiments for pre-modified titania with platinum are presently under study.

UV/Vis photocatalytic activity of Cu_2O/TiO_2 hybrid photocatalysts in this reaction system is enhanced by the combination of the synergistic effect of formed metallic copper and Cu_2O caused by the effect of Schottky barrier created between zero-valent copper and cuprous oxide (hindering charge carriers' recombination in cuprous oxide—its main shortcoming) [33,48] with a Z-scheme system (see Figure 7) as a type of mechanism of photogenerated carriers migration to form an efficient two-step charge separation system. Moreover, it should be pointed out that the proposed $Cu-Cu_2O-TiO_2$ nanostructure limits the problem of Cu_2O instability, i.e., self-oxidation by photo-generated holes (recombine with CB electrons from titania) and self-reduction by photo-generated electrons (which migrate to zero-valent copper).

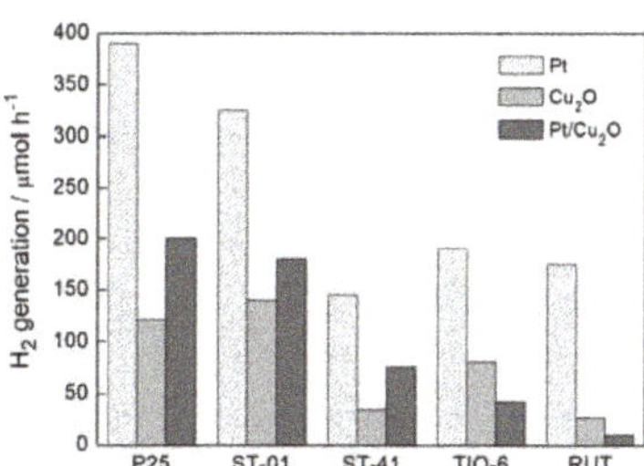

Figure 6. Comparison of UV/Vis photocatalytic efficiency of Cu_2O/TiO_2, $Pt/Cu_2O/TiO_2$ and Pt/TiO_2 samples in methanol dehydrogenation. TiO_2: P25, ST-01, ST-41, TIO-6, RUT.

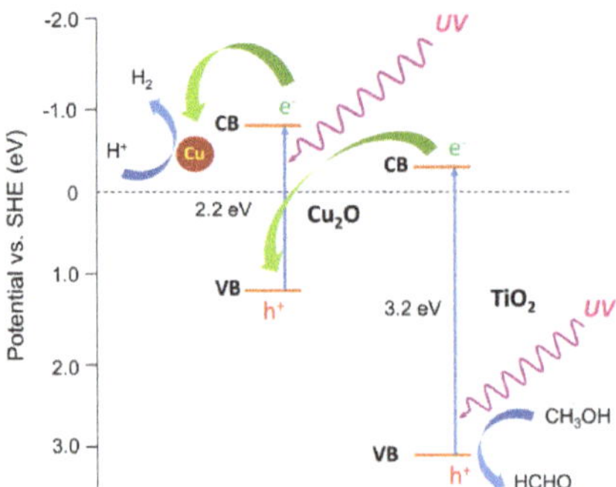

Figure 7. Two-step charge separation system (Z-scheme) with Cu-Cu$_2$O Schottky barrier as the mechanism for UV/Vis-induced methanol dehydrogenation with Cu (formed in-situ)/Cu$_2$O/TiO$_2$ photocatalysts.

3.2.2. UV/Vis-Induced Acetic Acid Oxidation

As in the previous reaction system, photocatalytic activity in the UV/Vis-induced acetic acid oxidation increased with an increase in Cu$_2$O content until 5 wt % (Figure 8). The exception was only for the TIO-6 sample, where the maximum of activity was found for 1 wt % of Cu$_2$O. A further increase of Cu$_2$O content was detrimental for the photocatalytic activity. Pure Cu$_2$O was almost inactive in this reaction because of a high recombination rate [49]. Therefore, either its high content in this hybrid photocatalyst or dark color (inner filter and shielding effects as discussed in Section 3.2.1) caused low photocatalytic efficiency under UV/vis irradiation. It should be pointed out that fine titania (ST-01) and mixed-phase titania (P25) are well known as highly active samples for this reaction (it is difficult to find more active titania photocatalysts, and probably only decahedral anatase particles (faceted anatase with two kinds of facets: eight {001} facets and two {001} facets) exhibited slightly higher activity than that of P25 [50]); and thus an increase in their activities by ca. 4–5 times by modification with small content of Cu$_2$O (Figure 9) is highly promising for other oxidation reactions and even the complete mineralization of organic pollutants.

For Cu$_2$O/TiO$_2$ samples with anatase as a dominant titania phase, the significant improvement of photocatalytic activity was achieved (Figure 9). The reaction efficiency was ca. 4–5 higher than that for corresponding bare TiO$_2$ regardless of particle size of titania, and catalytic activity (in the absence of irradiation) of Cu$_2$O/TiO$_2$ samples was negligible. It is important to mention that the preparation of these samples by physical mixing is not detrimental for overall photocatalytic performance of obtained heterojunction systems, which is really high. Significantly smaller improvement was observed only for rutile-based hybrid photocatalysts, in particular, for TIO-6. Figure 10 shows the results of the long photoactivity experiment for a Cu$_2$O/P25 sample considering the reusability of the photocatalyst and the comparison to the activity of bare P25. After 6 h of irradiation, an almost linear course of CO$_2$ liberation was still observed suggesting good photostability in this reaction system during continuous irradiation (the close reaction system with possible equilibrium between different forms of copper). However, the loss of photocatalytic activity of the recycled sample (losing the linear course) was observed for 2 h of irradiation. Fortunately, continued irradiation (2–6 h) resulted in stable photocatalytic activity. It was confirmed (by XRD analysis) that the content of the Cu$_2$O in recycled sample decreased, with simultaneous appearance of CuO and Cu (0) in comparison to fresh Cu$_2$O/P25 sample. Therefore, to extend the reusability of prepared samples, additional operations to strengthen the connection between these two components, e.g., annealing, or preparation of advanced nanostructures should be considered. For example, a core (Cu$_2$O)/shell (titania) nanostructure will be investigated in our future study, similarly to the reported Cu$_2$O/Au nanostructure with gold nanoparticles (NPs) deposited on Cu$_2$O nanowires [51].

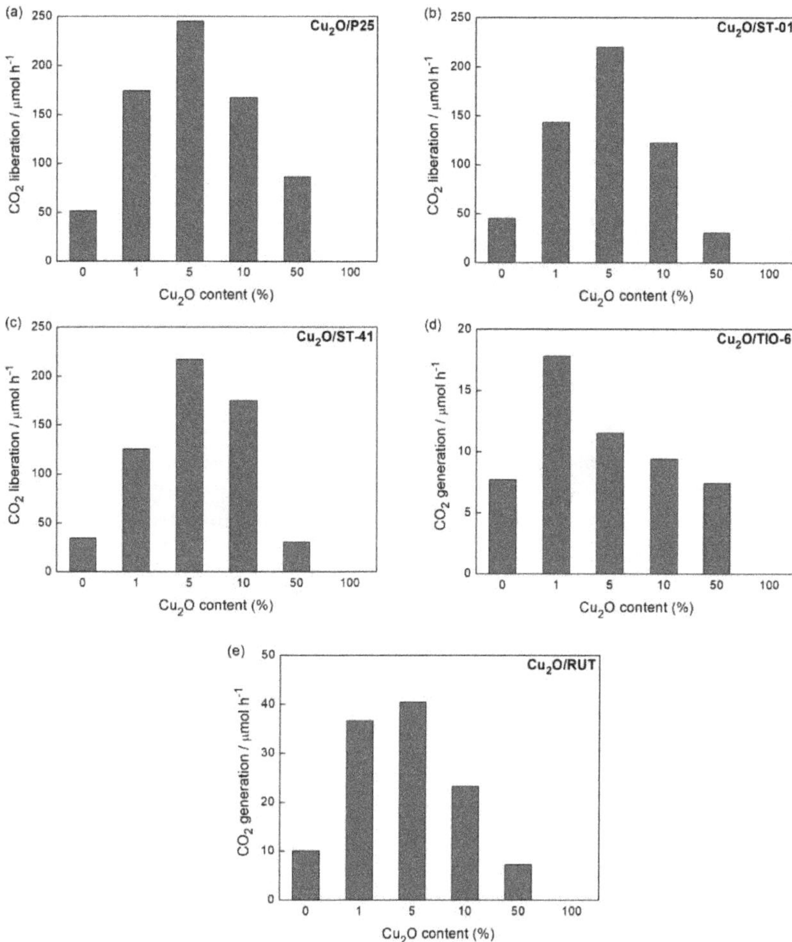

Figure 8. UV/Vis photocatalytic activity of samples: (**a**) $Cu_2O/P25$, (**b**) Cu_2O/ST-01, (**c**) Cu_2O/ST-41, (**d**) Cu_2O/TIO-6, (**e**) Cu_2O/RUT, prepared with corresponding Cu_2O content in acetic acid oxidation.

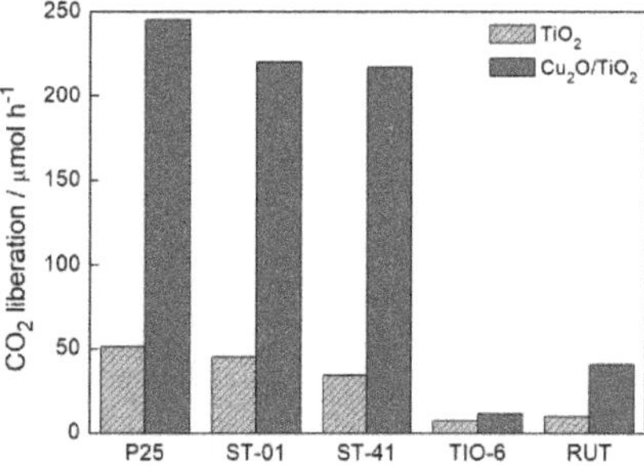

Figure 9. UV/Vis photocatalytic efficiency of Cu_2O/TiO_2 samples with different titania matrix and corresponding bare TiO_2 in acetic acid oxidation.

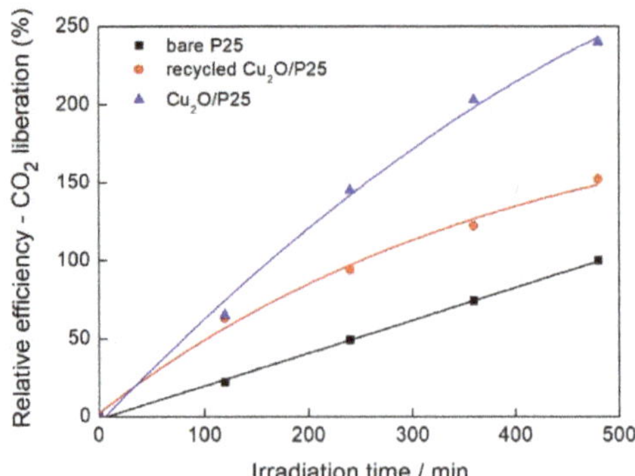

Figure 10. UV/Vis photocatalytic activity of Cu_2O/P25—the photostability study considering the reusability of Cu_2O/P25.

Two types of mechanism could be considered, similarly to H_2 system, i.e., Z-scheme and p-n heterojunction (type II), as shown in Figure 11. Under UV light irradiation, both Cu_2O and TiO_2 could be excited, and either photo-generated electrons in TiO_2 could recombine with photo-generated holes in the VB of Cu_2O or electrons in Cu_2O, and holes in TiO_2 could migrate to the CB of TiO_2 and VB of Cu_2O, respectively. The first mechanism seems to be preferential resulting in the generation of charges with stronger redox potential (more negative electrons and more positive holes). It is thought that photocatalytic activity for the oxidation reactions depends directly on the oxidation potential of holes, as recently reported for an oxygen activation study by M. Buchalska et al. [45]. The same study by M. Buchalska et al. proved that anatase was a stronger oxidant than rutile, due to the more positive position of the VB. Therefore, lower activities of rutile samples could be easily explained by less positive potential of the VB than that in anatase titania. Consequently, more negative potential of the CB in the rutile case may result in higher probability of type II heterojunction than Z-scheme, and thus not so high improvement of photocatalytic activity. Although, heterojunction II results in lower redox potential than the Z-scheme, the transfer process described above is thermodynamic favorable, and may result in the prolongation of the lifetime of excited electrons and holes, inducing higher quantum efficiency. Acetic acid is decomposed either by oxidative species such as $O_2{}^{\cdot-}$ and $OH^{\cdot}$, formed by the reaction of generated electrons with dissolved oxygen and by the reaction of generated holes from VB of TiO_2 with water, or directly by positive holes. It must be remembered that the lack of holes' consumption can be the reason for Cu_2O photocorrosion [52], and thus the proposed Z-scheme for anatase samples should be responsible for both high activity (strong redox ability) and stability.

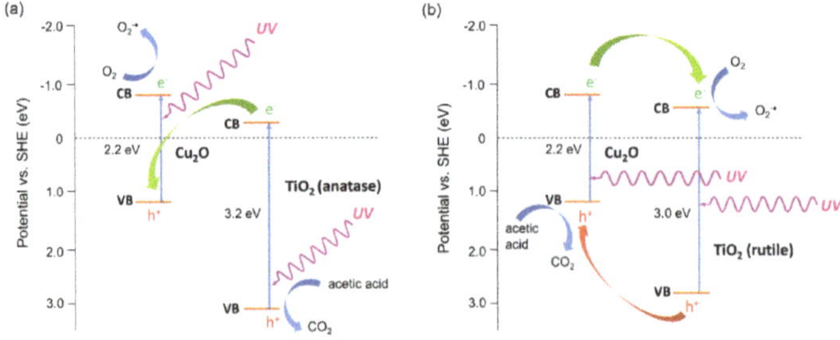

Figure 11. Proposed mechanisms of UV/Vis-induced acetic acid oxidation with Cu_2O/TiO_2 photocatalysts: Z-scheme for anatase samples (**a**) and p-n heterojunction (type II) for rutile sample (**b**).

3.2.3. Visible Light Photocatalytic Activity

Cu_2O as a 2.2 eV-band gap energy-semiconductor absorbs visible light. Therefore, one can expect that a hybrid system of Cu_2O with titania should be more active in the visible light than Cu_2O and titania alone. The results of photocatalytic activity for vis-induced 2-propanol oxidation (Figures 12 and 13) confirmed this expectation. Figure 14 shows the scheme of heterojunction system (Cu_2O/TiO_2) with visible light-activation of Cu_2O. The visible light-induced electron transfer between CB of Cu_2O and CB of titania should play the key role in the photocatalytic efficiency of this system. Similarly, as in the case of previous reaction systems 5 wt %-Cu_2O content resulted in the highest activity of Cu_2O/TiO_2 photocatalysts, independently of the titania matrix (Figure 12a–d), but with the exception of Cu_2O/RUT, where vis photocatalytic activity of bare RUT was higher than that in hybrid system (Figure 12e). The highest improvement of photocatalytic activity in relation to bare titania (ca. 6 times) was found for rutile sample: Cu_2O/TIO-6 (Figure 12d).

The crucial question is why the vis-induced photocatalytic properties of samples Cu_2O/TIO-6 and Cu_2O/RUT are so different. Densities of lattice defects (DEF; also known as electron traps (ETs)) equivalent to the concentration of Ti^{3+} were estimated for different types of titania in the earlier study [53]. The values of DEF were 50, 84, 38, 242 and 18 $\mu mol \cdot g^{-1}$ for P25, ST-01, ST-41, TIO-6 and RUT, respectively. The higher DEF of titania matrix favors the higher vis light-photocatalytic activity in the considered reaction system (Cu_2O/TIO-6), but the lowest DEF of RUT corresponds to no reaction rate improvement. The presence of Ti^{3+} ions can be important for the efficiency of the visible light-induced reaction on Cu_2O/TiO_2. Photogenerated electrons from Cu_2O can be captured by Ti^{4+} ions and thus, being reduced to Ti^{3+}. Ti^{3+} ions (with prolonged lifetime), participate in electron trapping resulting in retarded charge recombination. These observations are with agreement with the concept of Xiong et al. on the role of Ti^{3+} ions in Cu_2O/TiO_2 heterojunction system [23]. Moreover, the significant difference in enhancement factor between anatase and rutile hybrid samples of similar vis activity before modification (ST-01 and TIO-6, due to high content of DEF) could indicate that localization of CB of titania in respect to that of cuprous oxide is crucial. Therefore, higher proximity between cuprous oxide and rutile than that between cuprous oxide and anatase could facilitate an electron migration. Moreover, as M. Buchalska et al. suggested [45] the lower redox potential of the excited electron of rutile than that of anatase resulted in more efficient $O_2{}^{\cdot-}$ generation, and thus higher activity in reactions involving photo-excited electrons as the main mechanism pathway. Similarly, a higher activity of rutile than anatase was found for plasmonic photocatalysis by gold-modified titania, in which "hot" electron transfer from plasmonic gold NPs to CB of titania was proposed [54].

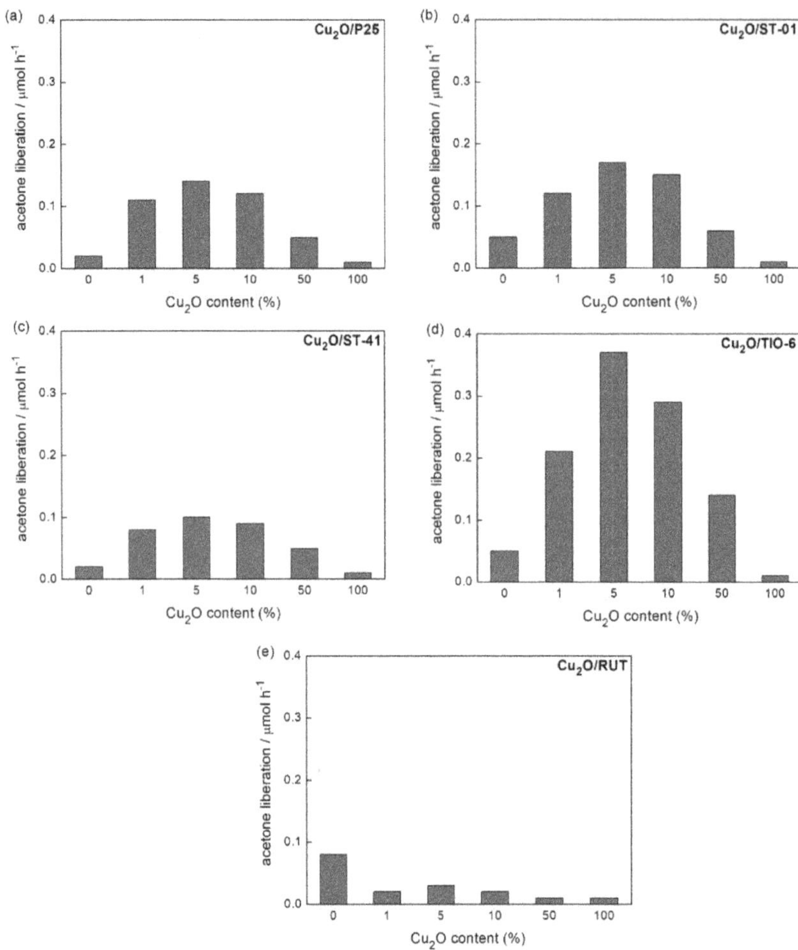

Figure 12. Visible light-photocatalytic activity of samples: (**a**) $Cu_2O/P25$, (**b**) $Cu_2O/ST-01$, (**c**) $Cu_2O/ST-41$, (**d**) $Cu_2O/TIO-6$, (**e**) Cu_2O/RUT, prepared with corresponding Cu_2O content in 2-propanol oxidation.

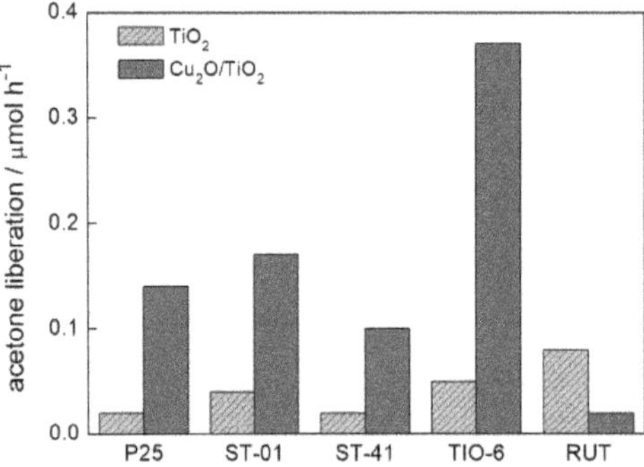

Figure 13. Visible light-photocatalytic efficiency of Cu_2O/TiO_2 samples with different titania matrix and corresponding bare TiO_2 in 2-propanol oxidation.

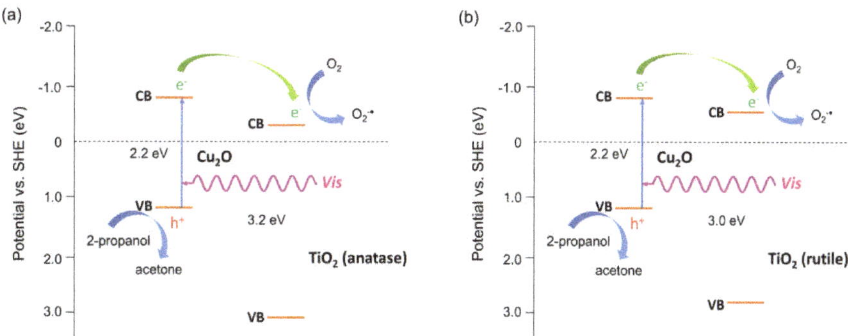

Figure 14. Schematic mechanism of heterojunction (type II) for vis-induced 2-propanol oxidation with Cu_2O/TiO_2 photocatalysts for anatase (**a**) and rutile (**b**).

3.3. Antimicrobial Properties of Cu_2O/TiO_2

At first, the bactericidal property of Cu_2O was investigated in the dark, under UV and visible light, and the results obtained are shown in Figure 15a. Cuprous oxide exhibited high bactericidal activity, and irradiation with UV and visible light slightly enhanced the intrinsic activity of Cu (I). It means that light irradiation could promote electron transfer between Cu and bacterial cells (Cu extracts electrons from bacteria, causing proteins denaturation [40]) and generation of reactive oxygen species (ROS) resulting in the cell's inactivation. It should be pointed out that the contact between Cu (I) and bacteria is essential for bacterial inactivation, in addition, surface Cu-ions are crucial for bactericidal property [55].

Bare titania exhibited bactericidal property under UV light irradiation, due to the generation of ROS, such as $^{\bullet}OH$, $O_2^{-\bullet}$ and H_2O_2, in which the activity seemed to arise from the particle size. In contrast, compared to the activity of anatase and without titania, bare rutile titania did not show significant enhancement under UV irradiation, which is not surprising because the photocatalytic activity of rutile is generally lower than that of anatase [54], as already discussed for acetic acid oxidation.

Under UV light irradiation, all Cu-modified anatase samples (Cu_2O/ST-01, $Cu_2O/P25$ and Cu_2O/ST-41) showed the enhancement of bactericidal activity. Qiu et al. have already reported that the bactericidal activity under visible light irradiation attributed to multi-electron reduction by electrons on Cu (II) in Cu_xO clusters which was transferred from the VB of titania by inter-facial charge transfer (IFCT), in contrast, Cu (I) in Cu_xO clusters showed anti-pathogen effect in the dark [55]. In this regard, it is proposed that similar mechanism could be responsible for enhanced UV-activity of Cu_2O/TiO_2 photocatalysts, i.e., IFCT from titania to Cu, as well as hindering of charge carriers' recombination (as discussed above). Interestingly, the dark activity of Cu_2O/ST-41 was higher than the visible one. It is probable that the contact between Cu (I) and bacteria could be affected, i.e., large titania ST-41 (crystal size = ca. 70 nm) could not cover Cu particles; on the other hand, small titania particles of ST-01 and P25 (ca. 8 and 21 nm, respectively) could cover NPs of cuprous oxide inhibiting the direct contact with bacteria and/or release of Cu ions from the surface of Cu (I). In the contrary, Cu (I)-modified rutile titania did not show the highest activity under UV irradiation, furthermore, the tendency of activity was different between two rutile samples (Cu_2O/TIO-6 and Cu_2O/RUT). In the case of Cu_2O/TIO-6, it is probable that direct bactericidal activity of Cu (I) in the dark might exceed the generation of ROS which was attributed to the activity of the multi-electron reduction on Cu by IFCT under UV and visible light irradiation, and vice versa in the case of Cu_2O/RUT.

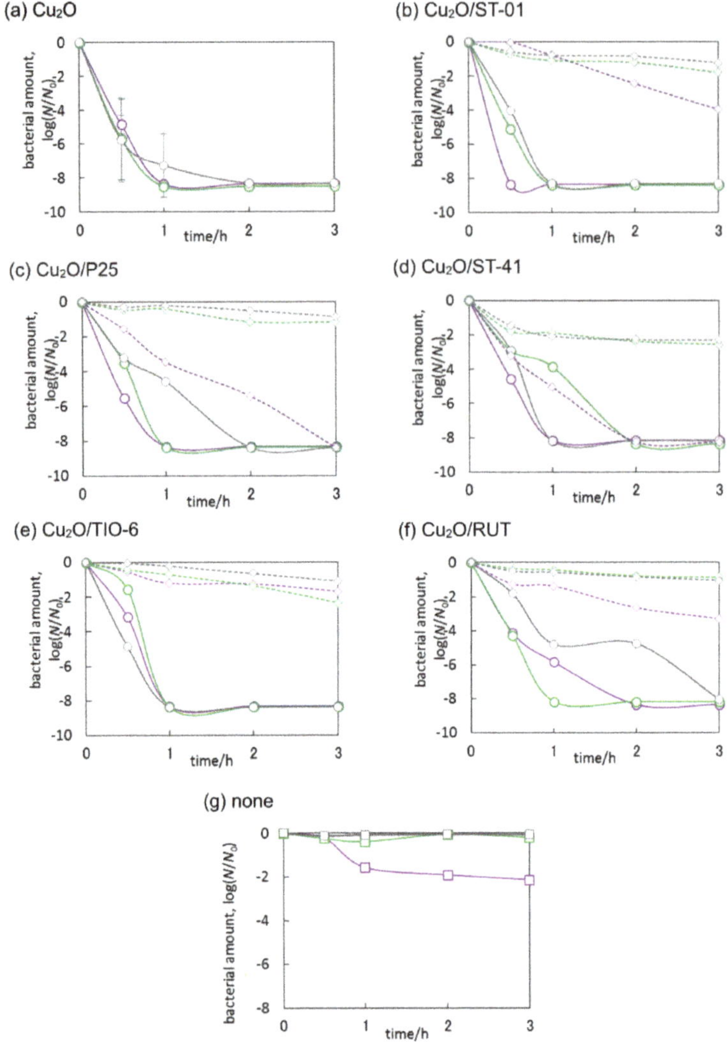

Figure 15. *E. coli* K12 survival shown as CFU/mL during inactivation of bacterial cells; (**a–f**) in the dark (grey symbols), under UV irradiation (300 < λ < 420 nm; violet symbols) and under vis irradiation (λ > 420 nm; green symbols) on bare (diamond, dashed line) and modified titania (circle, solid line), (**g**) *E. coli* K12 survival without titania in the dark, under UV and visible. Error bars (Cu (I) oxide) indicate the standard deviations calculated from two or three independent measurements.

The most active photocatalyst, i.e., Cu₂O/ST-01 was additionally tested under natural solar radiation, and data obtained are shown in Figure 14. Interestingly, no bactericidal activity was observed both in the absence of photocatalyst and for bare titania (ST-01), despite the fact that titania ST-01 showed some activity under UV irradiation (Figure 15b,g), possibly because the light intensity of solar radiation was quite low compared to an artificial source of light (xenon lamp), used in laboratory experiments (Figure 16). On the other hand, Cu₂O/ST-01 showed high activity under solar light and better than that in the dark (ca. one order of magnitude). It is thought that additionally to dark activity of Cu (I), and similarly to acetic acid oxidation, enhanced generation of reactive oxygen species could result in activity improvement (Z-scheme shown in Figure 11a).

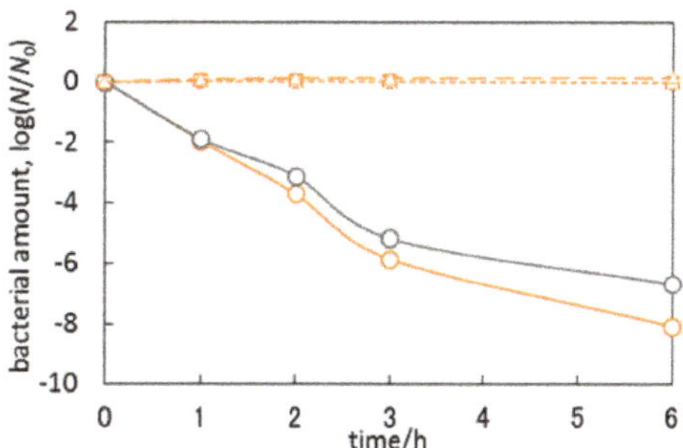

Figure 16. Bactericidal activities of Cu_2O/ST-01 (circle, solid line), bare titania ST-01 (square, dashed line) and reference experiments without any photocatalyst (triangle, dashed line) under solar light irradiation (orange) and in the dark (gray).

Concluding, it is clear that the bactericidal property of Cu_2O/TiO_2 originates mainly from the presence of cuprous ions, and photocatalytic activity only slightly enhanced the effect. Therefore, it is suggested that positively charged Cu (I) could both attract a negatively charged bacterial membrane (due to the presence of lipopolysaccharide) and inactivate cells by intrinsic activity of Cu (I).

In order to investigate deeper the antimicrobial effects of cuprous oxide/titania system, the fungicidal activities (*C. albicans*) were additionally studied, and data obtained are shown in Figure 17. It was found that copper (I) oxide remarkably suppressed fungal survival only under irradiation with UV light. The initial fungicidal rate was quite slow, and then accelerated after 1 h of irradiation. It is important to take into account the structure of fungal cells (yeast) and the surface charge of the cell wall. Although the electrostatic potential of *C. albicans* cells' surface is negative [56], their cell walls are rigid, they have a nuclear membrane and the size of cell is larger than that of bacteria. Therefore, despite Cu (I) oxide being easily in contact with cells, it could take a longer time to kill fungal cells than bacterial cells. It could be considered that the fungicidal mechanisms are similar to bacterial ones, and in addition, as reported by K. Danmek et al., Cu inhibits the activity of cellulase (in *Aspergillus melleus*), which could induce the inhibition of glycan decomposition and eventually the lack of nutrients [57]. The inactivation of *C. albicans* by cuprous oxide/titania was greatly promoted by UV light irradiation, and the activities in the dark were not so effective, unlike bactericidal activity in the dark. All Cu_2O-modified titania samples under UV irradiation reached detection limit within 1–2 h. Therefore, it is proposed that, in the case of fungi, although the influence of intrinsic activity of Cu (I) is slow, the effect of ROS on cell components might be fast, resulting in the difference of velocities between Cu_2O and Cu_2O/TiO_2. The activities under visible light (Cu (I) oxide and Cu_2O/ST-01) were almost the same as that in the dark suggesting that enhancement of antifungal activity was not only caused by possible formation of superoxide anion radicals (Figure 14). Accordingly, it is proposed that significant enhancement of activity under UV irradiation for Cu_2O/TiO_2 photocatalyst could result from a Z-scheme mechanism (Figure 11a) leading to either enhanced generation of hydroxyl radicals (by both holes from VB of titania and electrons from CB of cuprous oxide) or direct decomposition of fungal cells by photogenerated charge carriers. It will be clarified in our further studies by comparing ROS generated in UV, visible and dark conditions. Summarizing, it was found that fungicidal activities of cuprous oxide in the dark were promoted by modifying with titania, indicating that the activity was not derived from the sole activity of Cu_2O but also by heterojunction of Cu_2O and TiO_2.

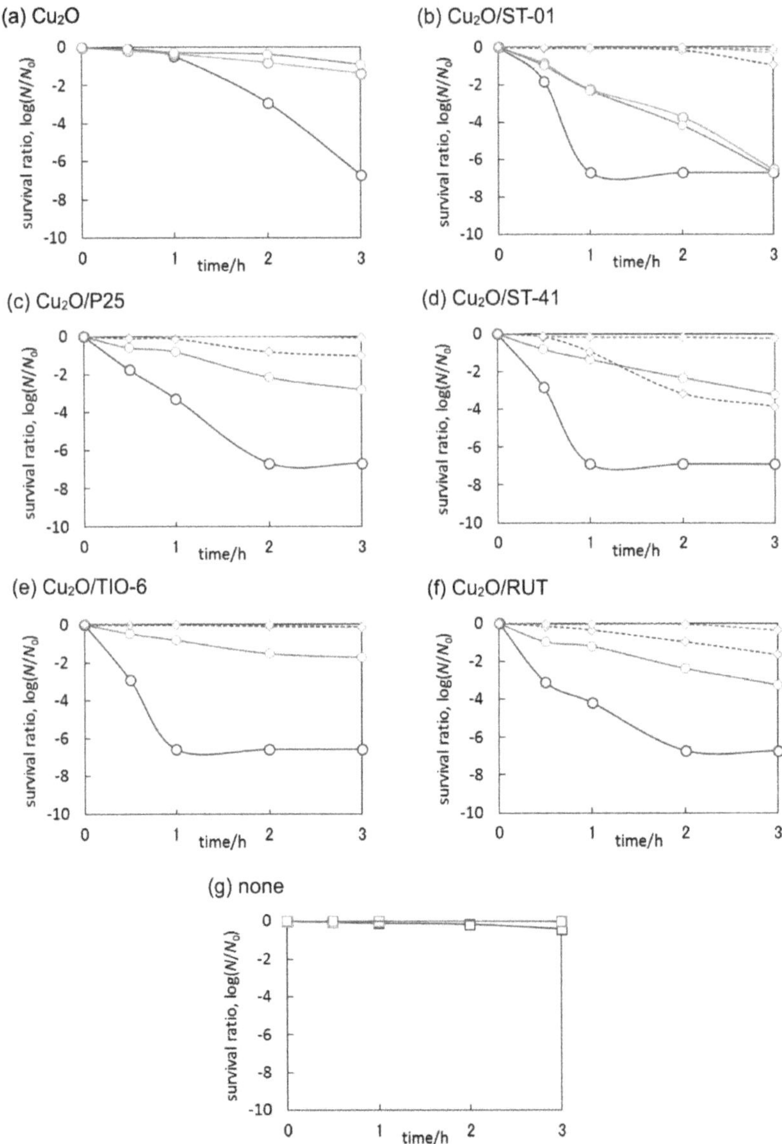

Figure 17. *C. albicans* survival shown as CFU/mL during inactivation of fungal cells; (**a–f**) in the dark (grey symbols), under UV irradiation (300 < λ < 420 nm; violet symbols) and under vis irradiation (λ > 420 nm; green symbols) on bare (diamond, dashed line) and modified titania (circle, solid line), (**g**) *E. coli* K12 survival without titania in the dark, under UV and visible.

4. Conclusions

In summary, a simple and low cost-method, realized by mixing of copper (I) oxide and titania, yields an efficient hybrid photocatalyst. By using a different titania matrix, one can adjust the both photocatalytic and antimicrobial properties of the resultant material. Considering the methanol dehydrogenation reaction, the enhanced efficiency of Cu_2O/TiO_2 photocatalysts originates from the combination of the Cu-Cu$_2$O Schottky barrier with a Z-scheme system. A large improvement of photocatalytic activity of copper (I) oxide-titania system in comparison to corresponding bare titanium(IV) oxides was found for UV/Vis-induced acetic acid oxidation, mainly for a titania matrix with anatase as a dominant phase. Taking into consideration oxidative reactions, Cu_2O/anatase is an example of a very efficient Z-scheme system, induced by UV/Vis irradiation, with a good perspective of application for solar systems dedicated for wastewater treatment, confirmed by good photostability, but the reusability of prepared photocatalysts needs further improvement. The mechanism of photocatalytic activity of rutile-based samples could be described as the type II heterojunction system. Furthermore, these two mechanistic variants, Z-scheme and heterojunction-type II, were also suggested for visible light-induced oxidation of 2-propanol for anatase and rutile-based samples, respectively. The photocatalytic efficiency in this system was correlated with the concentration of Ti^{3+} ions in a titania matrix (density of lattice defects)—the highest concentration of Ti^{3+} for TIO-6 means the highest vis-photocatalytic activity rate. Another important issue examined in this study was the antimicrobial property of Cu_2O/TiO_2 materials. All prepared samples possessed bactericidal and fungicidal properties, which were observed for UV, visible, solar irradiation, and even for dark conditions. It was concluded that antimicrobial activity depends not only on intrinsic properties of Cu_2O but also heterojunction between copper (I) oxide and titania.

Author Contributions: M.J. conceived, designed, performed the photocatalytic experiments and characterizations, interpreted the data and wrote the paper; M.E. performed antimicrobial tests and interpreted the data; K.W. performed XRD, DRS and photocatalytic experiments; D.Z. prepared samples and performed photocatalytic experiments; E.K. designed experiments, interpreted the data and corrected the manuscript. All authors read and approved the final manuscript.

Funding: This research was founded by Institute for Catalysis, Hokkaido University (HU) within the framework of research cluster on Plasmonic Photocatalysis, and Invitation Program for Foreign Professors (HU).

Acknowledgments: Authors thank Bunsho Ohtani for sound advice and unlimited access to laboratory equipment; Agata Markowska-Szczupak for fruitful discussions and fungi species (collection from West Pomeranian University of Technology); and Eloi Pourlier-Cucherat for the assistance in performing photocatalytic experiments. M.J. acknowledges Hokkaido University for the guest lecturer position (2016–2017).

Conflicts of Interest: The authors declare no conflict of interest.

References

1. Bahnemann, D.W. Photocatalytic water treatment: Solar energy applications. *Sol. Energy* **2004**, *77*, 445–459. [CrossRef]

2. Song, L.; Zhang, S.; Chen, B.; Ge, J.; Jia, X. Fabrication of ternary zinc cadmium sulfide photocatalysts with highly visible-light photocatalytic activity. *Catal. Commun.* **2010**, *11*, 387–390. [CrossRef]

3. Li, K.; Peng, B.S.; Peng, T.Y. Recent advances in heterogeneous photocatalytic CO_2 conversion to solar fuels. *ACS Catal.* **2016**, *6*, 7485–7527. [CrossRef]

4. Asahi, R.; Morikawa, T.; Ohwaki, T.; Aoki, K.; Taga, Y. Visible-light photocatalysis in nitrogen-doped titanium oxides. *Science* **2001**, *293*, 269–271. [CrossRef] [PubMed]

5. Zhou, N.; Lopez-Puente, V.; Wang, Q.; Polavarapu, L.; Pastoriza-Santos, I.; Xu, Q.H. Plasmon-enhanced light harvesting: Applications in enhanced photocatalysis, photodynamic therapy and photovoltaics. *RSC Adv.* **2015**, *5*, 29076–29097. [CrossRef]

6. Fujishima, M.; Takatori, H.; Tada, H. Interfacial chemical bonding effect on the photocatalytic activity of TiO_2–SiO_2 nanocoupling systems. *J. Colloid Interface Sci.* **2011**, *361*, 628–631. [CrossRef] [PubMed]

7. Martsinovich, N.; Troisi, A. How TiO_2 crystallographic surfaces influence charge injection rates from a chemisorbed dye sensitiser. *Phys. Chem. Chem. Phys.* **2012**, *14*, 13392–13401. [CrossRef] [PubMed]

8. Schneider, J.; Matsuoka, M.; Takeuchi, M.; Zhang, J.; Horiuchi, Y.; Anpo, M.; Bahnemann, D.W. Understanding TiO_2 photocatalysis: Mechanisms and materials. *Chem. Rev.* **2014**, *114*, 9919–9986. [CrossRef] [PubMed]

9. Kisch, H. Semiconductor photocatalysis—Mechanistic and synthetic aspects. *Angew. Chem. Int. Ed.* **2013**, *52*, 812–847. [CrossRef] [PubMed]

10. Bhanushali, S.; Ghosh, P.; Ganesh, A.; Cheng, W.L. 1D copper nanostructures: Progress, challenges and opportunities. *Small* **2015**, *11*, 1232–1252. [CrossRef] [PubMed]

11. Clarizia, L.; Spasiano, D.; Di Somma, I.; Marotta, R.; Andreozzi, R.; Dionysiou, D.D. Copper modified-TiO$_2$ catalysts for hydrogen generation through photoreforming of organics. A short review. *Int. J. Hydrogen Energy* **2014**, *39*, 16812–16831. [CrossRef]

12. Wei, Z.; Endo, M.; Wang, K.; Charbit, E.; Markowska-Szczupak, A.; Ohtani, B.; Kowalska, E. Noble metal-modified octahedral anatase titania particles with enhanced activity for decomposition of chemical and microbiological pollutants. *Chem. Eng. J.* **2017**, *318*, 121–134. [CrossRef] [PubMed]

13. Janczarek, M.; Wei, Z.; Endo, M.; Ohtani, B.; Kowalska, E. Silver- and copper-modified decahedral anatase titania particles as visible light-responsive plasmonic photocatalyst. *J. Photonics Energy* **2017**, *7*, 012008. [CrossRef]

14. Wei, Z.; Janczarek, M.; Endo, M.; Wang, K.; Balčytis, A.; Nitta, A.; Méndez-Medrano, G.; Colbeau-Justin, C.; Juodkazis, S.; Ohtani, B.; et al. Noble metal-modified faceted anatase titania photocatalysts: Octahedron versus decahedron. *Appl. Catal. B Environ.* **2018**, *237*, 547–587. [CrossRef]

15. Janczarek, M.; Kowalska, E. On the origin of enhanced photocatalytic activity of copper-modified titania in the oxidative reaction systems. *Catalysts* **2017**, *7*, 317. [CrossRef]

16. Mendez-Medrano, M.G.; Kowalska, E.; Lehoux, A.; Herissan, A.; Ohtani, B.; Bahena, D.; Briois, V.; Colbeau-Justin, C.; Rodrigues-Lopez, J.L.; Remita, H. Surface modification of TiO$_2$ with Ag nanoparticles and CuO nanoclusters for application in photocatalysis. *J. Phys. Chem. C* **2016**, *120*, 5143–5154. [CrossRef]

17. DeSario, P.A.; Pietron, J.J.; Brintlinger, T.H.; McEntee, M.; Parker, J.F.; Baturina, O.; Stroud, R.M.; Rolison, D.R. Oxidation-stable plasmonic copper nanoparticles in photocatalytic TiO$_2$ nanoarchitectures. *Nanoscale* **2017**, *9*, 11720–11729. [CrossRef] [PubMed]

18. Qiu, X.; Miyauchi, M.; Sunada, K.; Minoshima, M.; Liu, M.; Lu, Y.; Li, D.; Shimodaira, Y.; Hosogi, Y.; Kuroda, Y.; et al. Hybrid Cu$_x$O/TiO$_2$ nanocomposites as risk-reduction materials in indoor environments. *ACS Nano* **2012**, *6*, 1609–1618. [CrossRef] [PubMed]

19. Liu, L.M.; Yang, W.Y.; Li, Q.; Gao, S.A.; Shang, J.K. Synthesis of Cu$_2$O nanospheres decorated with TiO$_2$ nanoislands, their enhanced photoactivity and stability under visible light illumination, and their post-illumination catalytic memory. *ACS Appl. Mater. Interfaces* **2014**, *6*, 5629–5639. [CrossRef] [PubMed]

20. Chu, S.; Zheng, X.M.; Kong, F.; Wu, G.H.; Luo, L.L.; Guo, Y.; Liu, H.L.; Wang, Y.; Yu, H.X.; Zou, Z.G. Architecture of Cu$_2$O@TiO$_2$ core-shell heterojunction and photodegradation for 4-nitrophenol under simulated sunlight irradiation. *Mater. Chem. Phys.* **2011**, *129*, 1184–1188. [CrossRef]

21. Bessekhouad, Y.; Robert, D.; Weber, J.-V. Photocatalytic activity of Cu$_2$O/TiO$_2$, Bi$_2$O$_3$/TiO$_2$ and ZnMn$_2$O$_4$/TiO$_2$ heterojunctions. *Catal. Today* **2005**, *101*, 315–321. [CrossRef]

22. Yang, L.; Luo, S.; Li, Y.; Xiao, Y.; Kang, Q.; Cai, Q. High efficient photocatalytic degradation of p-nitrophenol on a unique Cu$_2$O/TiO$_2$ p-n heterojunction network catalyst. *Environ. Sci. Technol.* **2010**, *44*, 7641–7646. [CrossRef] [PubMed]

23. Xiong, L.B.; Yang, F.; Yan, L.L.; Yan, N.N.; Yang, X.; Qiu, M.Q.; Yu, Y. Bifunctional photocatalysis of TiO$_2$/Cu$_2$O composite under visible light: Ti^{3+} in organic pollutant degradation and water splitting. *J. Phys. Chem. Solids* **2011**, *72*, 1104–1109. [CrossRef]

24. Liu, L.; Gu, X.; Sun, C.; Li, H.; Deng, Y.; Gao, F.; Dong, L. In situ loading of ultra-small Cu$_2$O particles on TiO$_2$ nanosheets to enhance the visible-light photoactivity. *Nanoscale* **2012**, *4*, 6351–6359. [CrossRef] [PubMed]

25. Liu, L.; Yang, W.; Sun, W.; Li, Q.; Shang, J.K. Creation of Cu$_2$O@TiO$_2$ composite photocatalysts with p−n heterojunctions formed on exposed Cu$_2$O facets, their energy band alignment study, and their enhanced photocatalytic activity under illumination with visible light. *ACS Appl. Mater. Interfaces* **2015**, *7*, 1465–1476. [CrossRef] [PubMed]

26. Wang, J.Y.; Ji, G.B.; Liu, Y.S.; Gondal, M.A.; Chang, X.F. Cu$_2$O/TiO$_2$ heterostructure nanotube arrays prepared by an electrodeposition method exhibiting enhanced photocatalytic activity for CO$_2$ reduction to methanol. *Catal. Commun.* **2014**, *46*, 17–21. [CrossRef]

27. Wang, M.; Sun, L.; Cai, J.; Xie, K.; Lin, C. p–n Heterojunction photoelectrodes composed of Cu$_2$O-loaded TiO$_2$ nanotube arrays with enhanced photoelectrochemical and photoelectrocatalytic activities. *Energy Environ. Sci.* **2013**, *6*, 1211–1220. [CrossRef]

28. Zhang, J.; Zhu, H.; Zheng, S.; Pan, F.; Wang, T. TiO$_2$ Film/Cu$_2$O microgrid heterojunction with photocatalytic activity under solar light irradiation. *ACS Appl. Mater. Interfaces* **2009**, *1*, 2111–2114. [CrossRef] [PubMed]

29. Zhang, S.; Peng, B.; Yang, S.; Fang, Y.; Peng, F. The influence of the electrodeposition potential on the morphology of Cu_2O/TiO_2 nanotube arrays and their visible-light-driven photocatalytic activity for hydrogen evolution. *Int. J. Hydrogen Energy* **2013**, *38*, 13866–13871. [CrossRef]

30. Huang, L.; Peng, F.; Wang, H.; Yu, H.; Li, Z. Preparation and characterization of Cu_2O/TiO_2 nano–nano heterostructure photocatalysts. *Catal. Commun.* **2009**, *10*, 1839–1843. [CrossRef]

31. Huang, L.; Peng, F.; Yu, H.; Wang, H. Preparation of cuprous oxides with different sizes and their behaviors of adsorption, visible-light driven photocatalysis and photocorrosion. *Solid State Sci.* **2009**, *11*, 129–138. [CrossRef]

32. Singh, M.; Jampaiah, D.; Kandjani, A.E.; Sabri, Y.M.; Gaspera, E.D.; Reineck, P.; Judd, M.; Langley, J.; Cox, N.; Embden, J.; et al. Oxygen-deficient photostable Cu_2O for enhanced visible light photocatalytic activity. *Nanoscale* **2018**, *10*, 6039–6050. [CrossRef] [PubMed]

33. Li, L.; Xu, L.; Shi, W.; Guan, J. Facile preparation and size-dependent photocatalytic activity of Cu_2O nanocrystals modified titania for hydrogen evolution. *Int. J. Hydrogen Energy* **2013**, *38*, 816–822. [CrossRef]

34. Cheng, W.Y.; Yu, T.H.; Chao, K.J.; Lu, S.Y. Cu_2O-decorated mesoporous TiO_2 beads as a highly efficient photocatalyst for hydrogen production. *ChemCatChem* **2014**, *6*, 293–300. [CrossRef]

35. Xi, Z.H.; Li, C.J.; Zhang, L.; Xing, M.Y.; Zhang, J.L. Synergistic effect of Cu_2O/TiO_2 heterostructure nanoparticle and its high H_2 evolution activity. *Int. J. Hydrogen Energy* **2014**, *39*, 6345–6353. [CrossRef]

36. Li, Y.P.; Wang, B.W.; Liu, S.H.; Duan, X.F.; Hukey, Z.Y. Synthesis and characterization of Cu_2O/TiO_2 photocatalysts for H_2 evolution from aqueous solution with different scavengers. *Appl. Surf. Sci.* **2015**, *324*, 736–744. [CrossRef]

37. Kumar, D.P.; Reddy, N.L.; Kumari, M.M.; Srinivas, B.; Kumari, V.D.; Sreedhar, B.; Roddatis, V.; Bondarchuk, O.; Karthik, M.; Neppolian, B.; et al. Cu_2O-sensitized TiO_2 nanorods with nanocavities for highly efficient photocatalytic hydrogen production under solar irradiation. *Sol. Energy Mater. Sol. Cells* **2015**, *136*, 157–166. [CrossRef]

38. Lalitha, K.; Sadanandam, G.; Kumari, V.D.; Subrahmanyam, M.; Sreedhar, B.; Hebalkar, N.Y. Highly stabilized and finely dispersed Cu_2O/TiO_2: A promising visible sensitive photocatalyst for continuous production of hydrogen from glycerol:water mixtures. *J. Phys. Chem. C* **2010**, *114*, 22181–22189. [CrossRef]

39. Senevirathna, M.K.I.; Pitigala, P.K.D.P.; Tennakone, K. Water photoreduction with Cu_2O quantum dots on TiO_2 nano-particles. *J. Photochem. Photobiol. A Chem.* **2005**, *171*, 257–259. [CrossRef]

40. Sunada, K.; Minoshima, M.; Hashimoto, K. Highly efficient antiviral and antibacterial activities of solid-state cuprous compounds. *J. Hazard. Mater.* **2012**, *235*, 265–270. [CrossRef] [PubMed]

41. Deng, C.H.; Gong, J.L.; Zeng, G.M.; Zhang, P.; Song, B.; Zhang, X.G.; Liu, H.Y.; Huan, S.Y. Graphene sponge decorated with copper nanoparticles as a novel bactericidal filter for inactivation of *Escherichia coli*. *Chemosphere* **2017**, *184*, 347–357. [CrossRef] [PubMed]

42. Dankovich, T.A.; Smith, J.A. Incorporation of copper nanoparticles into paper for point-of-use water purification. *Water Res.* **2014**, *63*, 245–251. [CrossRef] [PubMed]

43. Yu, J.; Hai, Y.; Jaroniec, M. Photocatalytic hydrogen production over CuO-modified titania. *J. Colloid Interface Sci.* **2011**, *357*, 223–228. [CrossRef] [PubMed]

44. Dozzi, M.V.; Chiarello, G.L.; Pedroni, M.; Livraghi, S.; Giamello, E.; Selli, E. High photocatalytic hydrogen production on Cu (II) pre-grafted Pt/TiO_2. *Appl. Catal. B Environ.* **2017**, *209*, 417–428. [CrossRef]

45. Buchalska, M.; Kobielusz, M.; Matuszek, A.; Pacia, M.; Wojtyla, S.; Macyk, W. On oxygen activation at rutile- and anatase-TiO_2. *ACS Catal.* **2015**, *5*, 7424–7431. [CrossRef]

46. Scanlon, D.O.; Dunnill, C.W.; Buckeridge, J.; Shevlin, S.A.; Logsdail, A.J.; Woodley, S.M.; Catlow, C.R.A.; Powell, M.J.; Palgrave, R.G.; Parkin, I.P.; et al. Band alignment of rutile and anatase TiO_2. *Nat. Mater.* **2013**, *12*, 798–801. [CrossRef] [PubMed]

47. Hai, Z.; El Kolli, N.; Uribe, D.B.; Beaunier, P.; Jose-Yacaman, M.; Vigneron, J.; Etcheberry, A.; Sorgues, S.; Colbeau-Justin, C.; Chen, J.; et al. Modification of TiO_2 by bimetallic Au–Cu nanoparticles for wastewater treatment. *J. Mater. Chem. A* **2013**, *1*, 10829–10835. [CrossRef] [PubMed]

48. Li, Z.H.; Liu, J.W.; Wang, D.J.; Gao, Y.; Shen, J. $Cu_2O/Cu/TiO_2$ nanotube Ohmic heterojunction arrays with enhanced photocatalytic hydrogen production activity. *Int. J. Hydrogen Energy* **2012**, *37*, 6431–6437. [CrossRef]

49. Handoko, A.D.; Tang, J. Controllable proton and CO_2 photoreduction over Cu_2O with various morphologies. *Int. J. Hydrogen Energy* **2013**, *38*, 13017–13022. [CrossRef]

50. Janczarek, M.; Kowalska, E.; Ohtani, B. Decahedral-shaped anatase titania photocatalyst particles: Synthesis in a newly developed coaxial-flow gas-phase reactor. *Chem. Eng. J.* **2016**, *289*, 502–512. [CrossRef]

51. Pan, Y.; Polavarapu, L.; Gao, N.; Yuan, P.; Sow, C.H.; Hu, Q.H. Plasmon-enhanced photocatalytic properties of Cu_2O nanowire–Au nanoparticle assemblies. *Langmuir* **2012**, *28*, 12304–12310. [CrossRef] [PubMed]

52. Serpone, N.; Maruthamuthu, P.; Pichat, P.; Pelizzetti, E.; Hidaka, H. Exploiting the interparticle electron transfer process in the photocatalysed oxidation of phenol, 2-chlorophenol and pentachlorophenol: Chemical evidence for electron and hole transfer between coupled semiconductors. *J. Photochem. Photobiol. A Chem.* **1995**, *85*, 247–255. [CrossRef]

53. Prieto-Mahaney, O.O.; Murakami, N.; Abe, R.; Ohtani, B. Correlation between photocatalytic activities and structural and physical properties of titanium(IV) oxide powders. *Chem. Lett.* **2009**, *38*, 238–239. [CrossRef]

54. Kowalska, E.; Prieto Mahaney, O.O.; Abe, R.; Ohtani, B. Visible-light-induced photocatalysis through surface plasmon excitation of gold on titania surfaces. *Phys. Chem. Chem. Phys.* **2010**, *12*, 2344–2355. [CrossRef] [PubMed]

55. Baghriche, O.; Rtimi, S.; Pulgarin, C.; Sanjines, R.; Kiwi, J. Innovative TiO_2/Cu nanosurfaces inactivating bacteria in the minute range under low-intensity actinic light. *ACS Appl. Mater. Interfaces* **2012**, *4*, 5234–5240. [CrossRef] [PubMed]

56. Jones, L.; O'Shea, P. The electrostatic nature of the cell surface of Candida albicans: A role in adhesion. *Exp. Mycol.* **1994**, *18*, 111–120. [CrossRef]

57. Danmek, K.; Intawicha, P.; Thana, S.; Sorachakula, C.; Meijer, M.; Samson, R.A. Characterization of cellulase producing from Aspergillus melleus by solid state fermentation using maize crop residues. *Afr. J. Microbiol. Res.* **2014**, *8*, 2397–2404.

Article

Photocatalytic Performance of Cu$_x$O/TiO$_2$ Deposited by HiPIMS on Polyester under Visible Light LEDs: Oxidants, Ions Effect, and Reactive Oxygen Species Investigation

Hichem Zeghioud [1], Aymen Amine Assadi [2,*], Nabila Khellaf [3], Hayet Djelal [4], Abdeltif Amrane [2] and Sami Rtimi [5,*]

[1] Department of Process Engineering, Faculty of Engineering, Badji Mokhtar University, P.O. Box 12, 23000 Annaba, Algeria; hicheming@yahoo.fr

[2] Ecole Nationale Supérieure de Chimie de Rennes, Université de Rennes 1, CNRS, UMR 6226, Allée de Beaulieu, CS 50837, 35708 Rennes CEDEX 7, France; abdeltif.amrane@univ-rennes1.fr

[3] Laboratory of Organic Synthesis-Modeling and Optimization of Chemical Processes, Badji Mokhtar University, P.O. Box 12, 23000 Annaba, Algeria; khellafdaas@yahoo.fr

[4] Ecole des Métiers de l'Environnement, Campus de Ker Lann, 35170 Bruz, France; Hayet.Djelal@unilasalle.fr

[5] Ecole Polytechnique Fédérale de Lausanne, EPFL-STI-LTP, Station 12, CH-1015 Lausanne, Switzerland

* Correspondence: aymen.assadi@ensc-rennes.fr (A.A.A.); sami.rtimi@epfl.ch (S.R.)

Received: 29 December 2018; Accepted: 22 January 2019; Published: 29 January 2019

Abstract: In the present study, we propose a new photocatalytic interface prepared by high-power impulse magnetron sputtering (HiPIMS), and investigated for the degradation of Reactive Green 12 (RG12) as target contaminant under visible light light-emitting diodes (LEDs) illumination. The Cu$_x$O/TiO$_2$ nanoparticulate photocatalyst was sequentially sputtered on polyester (PES). The photocatalyst formulation was optimized by investigating the effect of different parameters such as the sputtering time of Cu$_x$O, the applied current, and the deposition mode (direct current magnetron sputtering, DCMS or HiPIMS). The results showed that the fastest RG12 degradation was obtained on Cu$_x$O/TiO$_2$ sample prepared at 40 A in HiPIMS mode. The better discoloration efficiency of 53.4% within 360 min was found in 4 mg/L of RG12 initial concentration and 0.05% Cu$_{wt}$/PES$_{wt}$ as determined by X-ray fluorescence. All the prepared samples contained a TiO$_2$ under-layer with 0.02% Ti$_{wt}$/PES$_{wt}$. By transmission electron microscopy (TEM), both layers were seen uniformly distributed on the PES fibers. The effect of the surface area to volume (dye volume) ratio (SA/V) on the photocatalytic efficiency was also investigated for the discoloration of 4 mg/L RG12. The effect of the presence of different chemicals (scavengers, oxidant or mineral pollution or salts) in the photocatalytic medium was studied. The optimization of the amount of added hydrogen peroxide (H$_2$O$_2$) and potassium persulfate (K$_2$S$_2$O$_8$) was also investigated in detail. Both, H$_2$O$_2$ and K$_2$S$_2$O$_8$ drastically affected the discoloration efficiency up to 7 and 6 times in reaction rate constants, respectively. Nevertheless, the presence of Cu (metallic nanoparticles) and NaCl salt inhibited the reaction rate of RG12 discoloration by about 4 and 2 times, respectively. Moreover, the systematic study of reactive oxygen species' (ROS) contribution was also explored with the help of iso-propanol, methanol, and potassium dichromate as •OH radicals, holes (h$^+$), and superoxide ion-scavengers, respectively. Scavenging results showed that O$_2$$^-$ played a primary role in RG12 removal; however, •OH radicals' and photo-generated holes' (h$^+$) contributions were minimal. The Cu$_x$O/TiO$_2$ photocatalyst was found to have a good reusability and stability up to 21 cycles. Ions' release was quantified by means of inductively coupled plasma mass spectrometry (ICP-MS) showing low Cu-ions' release.

Keywords: photocatalysis; reactive green 12; Cu$_x$O/TiO$_2$; polyester; HiPIMS; visible light LEDs

---

1. Introduction

The textile industry is one of the largest consumers of water on our planet, and the second most polluting industry after the oil and gas industry [1,2]. Textile effluent containing synthetic dyes have toxic and carcinogenic compounds, which are stable and non-biodegradable due to the high molecular weight, the presence of azo bonds and amide groups in the molecular structure. Many of the chemical and physical treatment processes have shown insufficient results toward the degradation of theses pollutants. Recently, heterogeneous photocatalysis is one of the advanced oxidation processes that increasingly interests researchers. Photocatalysis was seen to transform/mineralize these synthetic dyes to lesser harmful by-products before their discharge into the environment. In the last decade, numerous works have been mostly focused on the designing and the development of new photocatalytic materials [3–8]. In this direction, different methods and preparations were investigated [3]. Sol–gel [4], hydrothermal [5], combined sol–gel/hydrothermal [6], chemical vapor deposition (CVD) [7], liquid phase deposition (LPD) methods [8] among many others were reported in the literature. In the aim of understanding the relationship between physico-chemical proprieties of photocatalyst and photocatalytic performances, various strategies have been adopted. Many studies investigated the effect of (1) semiconductor type such as TiO_2, ZnO, and SnO_2 [9,10]; (2) the number (mono-, bi-, or tri-doping) and the ratio of doping agents (N, P, Fe, Ag, etc.) [11–14]; (3) the support type such as polyester, cellulose, polypropylene, or polystyrene [15–17]; and (4) the substrate functional groups before the catalyst deposition [18,19]. It had been widely reported that band gap energy, surface area, particles size, and chemical stability were the most important parameters controlling the reactivity of a photocatalytic material [20–22]. TiO_2 was reported to be the most suitable photocatalyst because of its high stability, low toxicity, low cost, chemical inertness, wide band gaps, and resistance to photo-corrosion [23–25]. Conventional direct current magnetron sputtering (DCMS) and high-power impulse magnetron sputtering (HiPIMS) have been used during the last decade to prepare photoactive thin films [26,27]. These films were reported to exhibit redox catalysis leading to bacterial inactivation [19,24,27], organic dyes degradation [25], antifungal [28], corrosion resistance [29], and self-cleaning [30]. Sequentially sputtered TiO_2/Cu showed bacterial inactivation in the minute range [31,32]. Co-sputtered TiO_2/Cu_xO using HiPIMS was reported to lead to compact photoactive films showing reduced ions' release [31]. The main goals of the present study are (i) to explore the HiPIMS deposition of Cu_xO on TiO_2 under-layer; (ii) to optimize the deposition parameters leading to stable thin film materials showing fast degradation of a toxic textile dye (Reactive Green 12) as target hazardous compound; (iii) to use the visible light light-emitting diodes (LEDs) system as efficient and economic light source; and (iv) to test the effect of the presence of some oxidant, mineral pollutant, or salts on the performance of the photoactive material. This latter goal is a step further to mimic real textile industry wastewater effluents normally presenting salts and oxidative agents. The photo-generation and the contribution of reactive oxygen species (ROS) are investigated in detail.

2. Experimental

2.1. Materials

Hydrogen peroxide (35 wt %, Merck KGaA, Darmstadt, Germany), potassium dichromate (>99 wt %, Carlo Erba Reagents S.A.S., Chaussée du Vexin, France), 2-propanol (>99.5%, Merck KGaA, Darmstadt, Germany), potassium peroxydisulfate ($\geq$90 wt %, Merck KGaA, Darmstadt, Germany), chloroform (>99.97%, Acros-Organics, Thermo Fisher Scientific, Geel, Belgium), methanol (99.99%, Thermo Fisher Scientific, Geel, Belgium), copper (98.5 wt %, Merck KGaA, Darmstadt, Germany), and sodium chloride (99.5 wt %, Acros-Organics, Thermo Fisher Scientific, Geel, Belgium) were used. Reactive Green 12 dye (>99%, MW = 1837 g mol^{-1}) procured from the textile manufacture of Constantine (Algeria) was used as received; the aqueous solutions were prepared using MilliQ water with a resistance of 15.0 MΩ·cm. The chemical structure and properties of Reactive Green 12 (RG12) dye were recently reported [18].

2.2. Catalyst Preparation

The Cu_xO/TiO_2 coatings were sputtered using reactive HiPIMS process. Ti and Cu targets were purchased from Lesker (Kurt J. Lesker Company Ltd., East Sussex, UK) (99.99% pure). The sputtering chamber was operated at a high vacuum with a residual pressure less than 4×10^{-5} Pa. This chamber was equipped with two confocal targets 5 cm in diameter. A HiP3 5KW Solvix generator (Advanced Energy, Fort Collins, CO, USA) was used for the HiPIMS deposition and was operated at an average power of 100 W (5 W·cm^{-2}) with a pulse-width of 50 μs and a frequency of 1000 Hz. A TiO_2 under-layer was sputtered (for 1 min) before the deposition of Cu_xO to reduce the voids of polyester and to permit a high dispersion of Cu species on the surface [33]. The target-to-substrate distance was fixed at 10 cm in order to obtain homogeneous and adhesive Cu_xO films. The sample holder was rotating at a speed of 18 rpm.

Table 1 shows a summary about the prepared catalysts used in this study. Different current intensities were applied to the Cu-sputtering target—(1) HiPIMS mode (20 A, 40 A and 80 A) and (2) direct current mode (direct current magnetron sputtering, DCMS) at 300 mA. The Cu-deposition times were 5, 10, 20, and 100 s. The obtained photocatalytic thin layers were used in the degradation of 4 mg/L of RG12 solution, as shown in Figure 1.

Table 1. Summary of the prepared catalysts used in this study.

Catalyst	Sputtering Mode	Sputtering Time of Copper Upper Layer (s) *	Sputtering Power
Cu_xO/TiO_2	DCMS/DCMS	5, 10, 20, and 100	0.5 A/0.3 A
Cu_xO/TiO_2	HiPIMS/DCMS	5, 10, 20, and 100	20 A/0.3 A
Cu_xO/TiO_2	HiPIMS/DCMS	5, 10, 20, and 100	40 A/0.3 A
Cu_xO/TiO_2	HiPIMS/DCMS	5, 10, 20, and 100	80 A/0.3 A

* The TiO_2 under-layer was sputtered for 1 min.

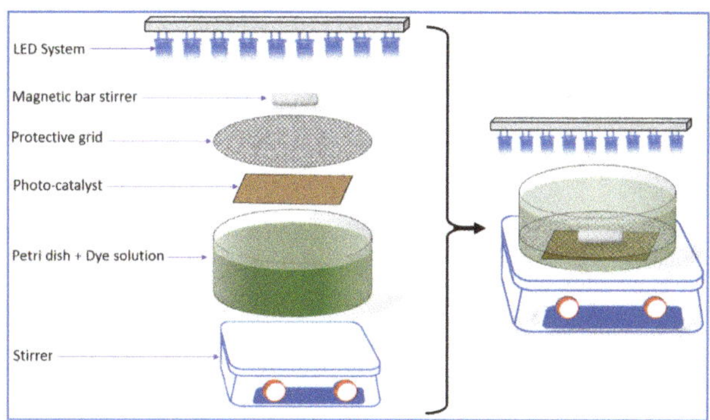

Figure 1. Photocatalytic apparatus for Reactive Green 12 (RG12) degradation.

2.3. Characterization of Materials

The UV–Vis spectra of all samples were recorded using a Varian Cary®50 UV–Vis spectrophotometer (Agilent, Les Ulis, France), where the spectra were obtained at the wavelength range of 200–800 nm (λ_{max} = 615 nm).

Transmission electron microscopy (TEM) of the sputtered polyester (PES) fabrics was carried out using a FEI Tecnai Osiris (Thermo Fisher Scientific, Hillsboro, OR, USA) operated at 200 kV. The spot size was set to 5 μm, dwell time 50 μs, and the real time of 600 s. The samples were embedded in epoxy resin (Embed 812) and cross-sectioned with an ultra-microtome (Ultracut E) up to thin layers of 80–100 nm thick. These thin layers were then placed on TEM holey carbon grid in order to image the samples.

X-ray diffraction (XRD) INEL EQUINOX instrument (INEL, Stratham, NH, USA), power 3.5 kW and coupled with a CPS120detector (INEL, Stratham, NH, USA) was used to register peaks from 2 to 120 θ.

2.4. Photocatalytic Experiments

The photocatalytic activity of the synthesized photocatalyst was evaluated by following the degradation of 15 mL of the RG12 dye solution in petri dish as a reactor under magnetic stirring (see Figure 1). The reaction system included an ultraviolet light-emitting diodes (LEDs) system (visible light LEDs, Innolux, Santa Clara, CA, USA) as an irradiation source (λ = 420 nm) with a measured intensity of 1 mW/cm². The initial RG12 solution was stirred in the dark for 30 min to reach adsorption–desorption equilibrium before the LEDs irradiation was ON. The concentration of dye solution samples (V = 3 mL) was analyzed using a UV–Vis spectrophotometer (Agilent, Les Ulis, France) at regular time intervals. The protective grid is made of stainless steel and is used to protect the catalyst from the possible mechanical damage caused by the stirrer.

The RG12 discoloration efficiency (%) of the material was evaluated by the following Equation (1)

$$n(\%) = \left(\frac{A_0 - A_t}{A_0}\right) \times 100 \tag{1}$$

where A_0 and A_t are the initial intensity of absorbance peak at λ_{max} (λ_{max} = 615 nm) and the intensity of absorbance peak at time t in UV–Vis spectra of RG12, respectively.

The coated fabrics were also tested for stability by testing their recycling performance. The ions released from the fabrics were quantified using inductively coupled plasma mass spectrometry (ICP-MS) using Finnigan™, Element2 high-performance high-resolution ICPMS model (Zug, Switzerland). The ICP-MS resolution was of 1.2×10^5 cps/ppb with a detection limit of 0.2 ng/L. Clean Teflon bottles were used to prepare the calibration standards through successive dilutions of 0.1 g L^{-1} ICPMS stock solutions (TechLab, Metz, France). The washing solution from the samples were digested with nitric acid 69% (1:1 HNO$_3$ + H$_2$O) to remove the organics and to guarantee that there were no remaining ions adhered to the reactor wall.

3. Results and Discussion

3.1. Effect of the Photocatalyst Preparation Parameters on the RG12 Discoloration and the Microstructure

The photocatalytic degradation of RG12 under the LEDs is shown in Figure 2. We noted that when applying a current intensity of 40 A to the target, the resulting thin film showed the fastest RG12 degradation followed by HiPIMS at 20 A, DCMS (300 mA), and HiPIMS at 80 A with 100 s as deposit time.

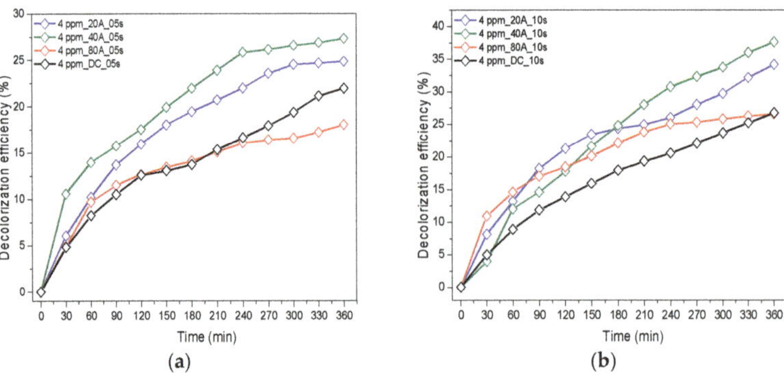

Figure 2. *Cont.*

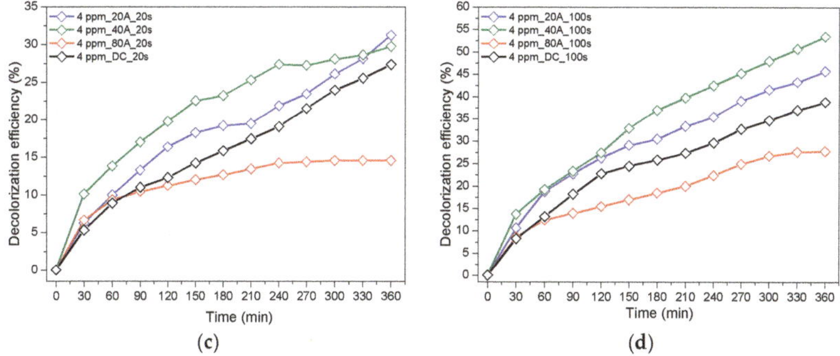

(c) (d)

Figure 2. Photocatalytic degradation of RG12 with different current intensity used in photocatalyst preparation (initial pollutant concentration 4 mg/L) on (**a**) samples sputtered for 5 s on polyester (PES); (**b**) samples sputtered for 10 s; (**c**) samples sputtered for 20 s; and (**d**) samples sputtered for 100 s.

From another hand, at this value of current intensity (40 A), increasing the sputtering time led to an increase in the photocatalytic discoloration efficiency of RG12, where 27.3%, 33.1%, 37.6%, and 53.4% dye elimination after 360 min under irradiation was observed with 05, 10, 20, and 100 s deposition time, respectively, as shown below in Figure 2. The TiO_2 under-layer did not show any photocatalytic activity (3–6% RG12 removal). This is in accordance with previous results found for methylene blue [25]. This can be attributed to the low amount of TiO_2 (and active sites) available for RG12.

This can be attributed to the optimal mass-to-volume ratio of the HiPIMS deposited film at 40 A. It has been reported that small-sized nanoparticles induce favorable photocatalytic bacterial inactivation kinetics due to the large surface area per unit mass [24]. The distribution of Cu_xO nanoparticles on the TiO_2 under-layer on the polyester was found to be uniform and did not induce cracks on the substrate. Figure 3 shows the TEM imaging of the sputtered layers (HiPIMS at 40 A) on polyester (PES). The TiO_2 under was sputtered to reduce the porosity of the PES leading to the continuous distribution of the Cu_xO upper layer [31,33].

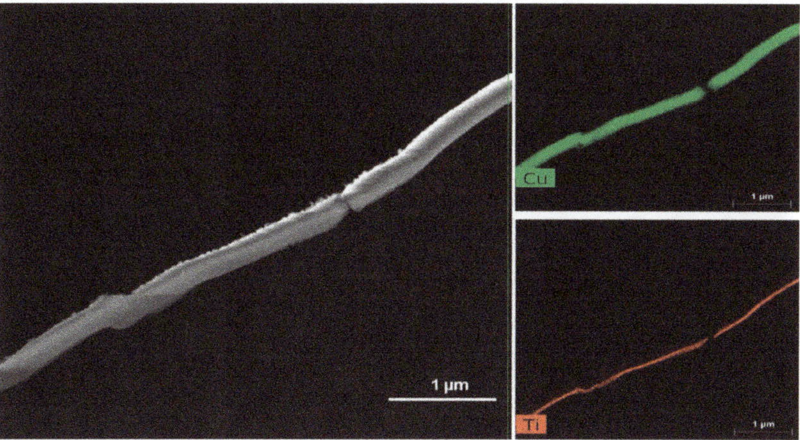

Figure 3. TEM imaging of Cu_xO/TiO_2 deposited by high-power impulse magnetron sputtering (HiPIMS) at 40 A.

The charge transfer between the TiO_2 and the Cu_xO at the surface and the organic pollutant depends on the diffusion length of the photo-generated charges at the interface of the film under light irradiation. The charge transfer/diffusion is also a function of the TiO_2 and of the Cu-particles' size and shape as previously reported [34,35].

X-ray diffraction of the sputtered Cu_xO/TiO_2 on PES (sputtered for 100 s at 40 A) showed the absence of sharp peaks that can be attributed to the Cu_xO clusters. This reflected the low crystallinity of the deposited Cu clusters. The absence of clear crystal phase could be attributed to the short sputtering time leading to the formation of very small Cu nanoparticles/clusters. The low Cu and Ti loadings led to ~80 nm thin film on the porous PES. The diffractogram showed a high noise-to-signal ratio due to the low amount of Cu species on the PES.

3.2. Effect of key Parameters: Pollutant Concentration and Surface-Area to Volume Ratio

Figure 4 shows the photocatalytic RG12 discoloration when using different initial dye concentrations. It is readily seen from Figure 4 that the discoloration is faster for lower initial concentrations, with optimal kinetics for 4 mg/L. This can be explained by the unavailability of active site for the adsorption of all dye molecules and their degradation at high initial concentrations [25]. Furthermore, the increase in dye concentration led to an increase in medium opacity and then less permeability to the applied light, which reduced the photo-generated reactive species important for the photocatalytic process [25,36]. It has been recently reported that porous Cu_2O nanostructures were also able to adsorb ions such as radioactive iodine issued from nuclear fission [37].

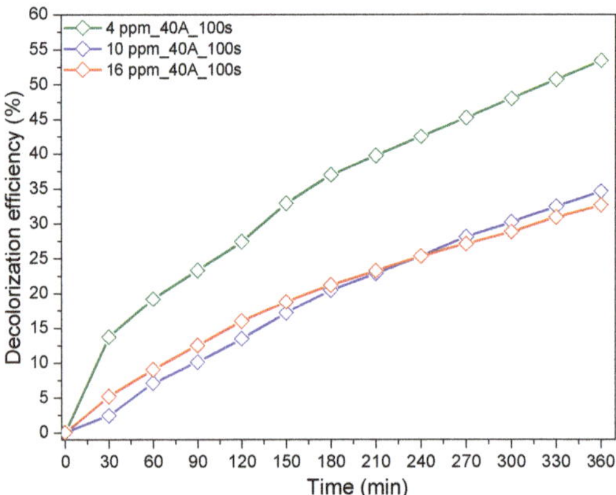

Figure 4. Photocatalytic degradation of RG12 with different initial concentrations (current intensity, 40 A and sputtering time, 100 s).

In order to investigate the effect of the photocatalyst surface (in other words, the available catalytic active sites), the effect of increasing surface area to volume (dye volume) ratio (SA/V) on the photocatalytic efficiency was studied using higher ratios for the discoloration of 4 mg/L RG12. We noted that the used photocatalytic surface in the previous experiment was fixed to 16 cm^2, which gave a surface area to volume ratio (SA/V) of 1.06 cm^{-1}. The results are shown in Figure 5. It was readily seen that increasing SA/V increased the discoloration efficiency. This can be attributed to the increase in the number of active sites available for RG12 molecules and intermediate by-products' adsorption and degradation. A ratio of 1.06 showed 53.4% of discoloration efficiency; however, a ratio of 2.66 cm^{-1} led to 90.1% under visible light LEDs irradiation for 360 min. Similar results were reported by Huang et al. for the methyl orange discoloration on Pt-modified TiO_2 supported on natural zeolites [38].

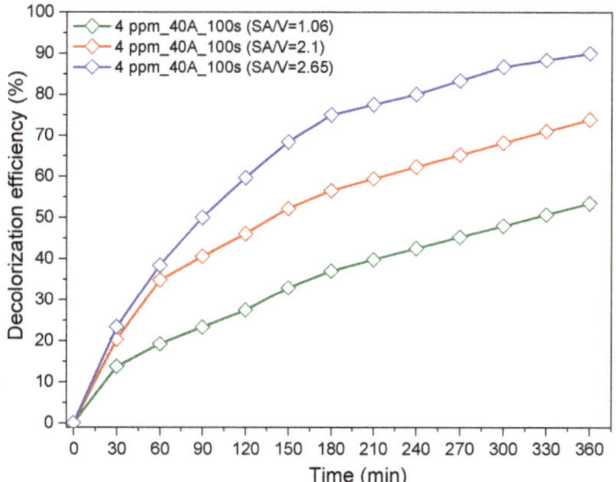

Figure 5. Photocatalytic degradation of RG12 with different catalyst surface area to volume ratio (initial concentrations, 4 mg/L; current intensity, 40 A; and sputtering time, 100 s).

3.3. Implication of Radicals and ROS Species within RG12 Discoloration

The contribution of individual reactive oxidizing species ($^{\bullet}OH$, $O_2^{\bullet-}$, and h^+) in the RG12 degradation was studied using selective scavengers (MeOH and iso-propanol for $^{\bullet}OH$, potassium dichromate for $O_2^{\bullet-}$, and $CHCl_3$ and MeOH for electronic holes). The choice of selective scavenger was based on the second-order kinetic rate constant of the reaction between the ROS and each scavenger as shown in Supplementary Materials Table S1.

Figure 6a shows that the presence of MeOH (holes scavenger) in reaction medium enhanced the discoloration efficiency of the reactive dye. Furthermore, increasing the MeOH amount led to the acceleration of the photocatalytic RG12 abatement until an optimum value of alcohol of 50 mmol/L leading to 74.3% of discoloration efficiency after 360 min of irradiation.

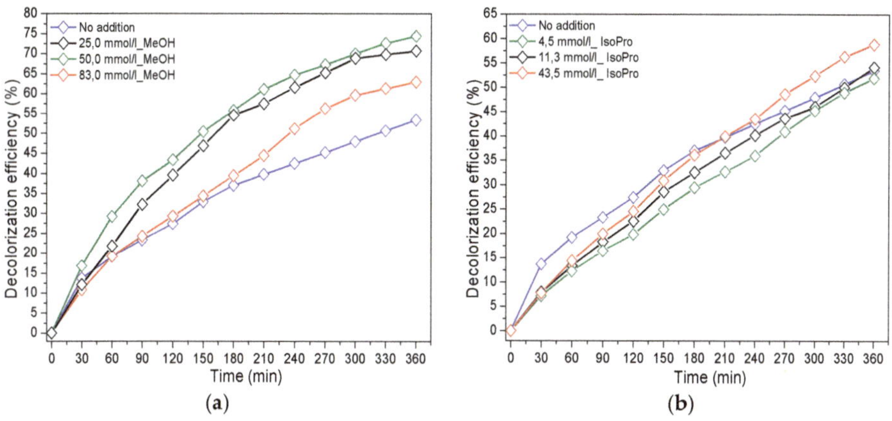

Figure 6. *Cont.*

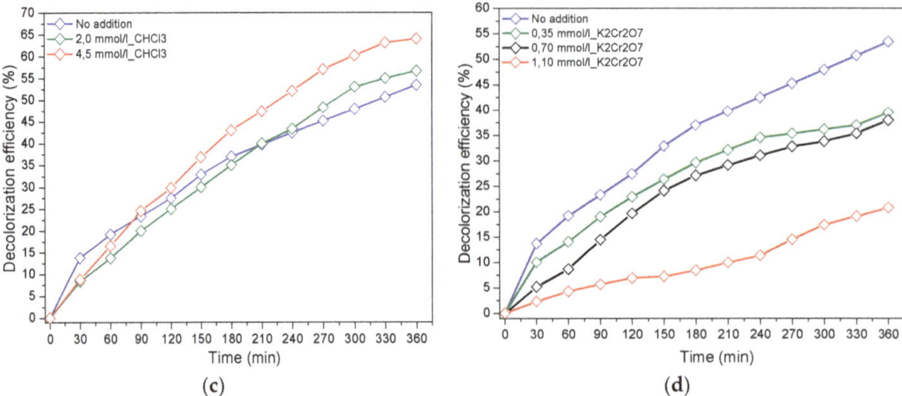

Figure 6. Photocatalytic degradation of RG12 in the presence of (**a**) methanol; (**b**) iso-propanol; (**c**) chloroform; and (**d**) potassium dichromate. (Initial RG12 concentration = 4 mg/L; sample, HiPIMS at 40 A; and sputtering time, 100 s.)

On the other hand, as shown in Figure 6b, the presence of iso-propanol in the reaction medium slightly enhanced the photocatalytic RG12 discoloration, where 58.8% of RG12 was destroyed with 43.5 mmol/L of alcohol after 360 min compared to 53.4% without scavenger addition.

Figure 6a,b shows that the presence of alcohol (MeOH or iso-propanol) slightly enhanced the photocatalytic reaction rate, which can be explained by the inhibitory effect played by hydroxyl radicals in the degradation mechanism of RG12. Similar results were found by Rong and Sun in their work reported on the photocatalytic degradation of triallyl isocyanurate (TAIC) [39]. The additional effect in the inhibition of discoloration efficiency in the presence of MeOH compared to iso-propanol can be attributed to the holes' effect because of the high reactivity of iso-propanol with the hydroxyl radicals ($k_{\text{iso-prop},{}^{\bullet}\text{OH}} = 1.9 \times 10^9\ \text{M}^{-1}\ \text{S}^{-1}$) [40] compared to the MeOH ($k_{\text{MeOH},{}^{\bullet}\text{OH}} = 9.7 \times 10^8\ \text{M}^{-1}\ \text{S}^{-1}$) [41].

It is worth to mention at this level that Di Valentin and Fittipaldi [42] studied the photo-generated hole scavenging using different organic adsorbates on TiO_2. They established a scale of scavenging power showing glycerol > tert-butanol > iso-propanol > methanol > formic acid. This suggested that methanol could exhibit electronic-holes scavenging, but negligible compared to the ${}^{\bullet}$OH-radical scavenging. Furthermore, Shen and Henderson [43] reported on the molecular and dissociative forms of methanol on TiO_2 surface for holes scavenging. They showed that methoxy (dissociative form of methanol) effectively scavenged the photo-generated holes, and not methanol itself. This did not allow an unequivocal separation between ${}^{\bullet}$OH-radical and h^+ scavenging when using methanol.

Figure 6c shows that the presence of 4.5 mmol/L of chloroform slightly improved the discoloration efficiency of RG12 by 11% after 360 min under light. This enhancement can be attributed to the holes-scavenging characters of Cl ions generated from the chloroform decomposition/dissolution.

Oxygen superoxide anion ($O_2{}^{\bullet-}$) is one of the very important radicals studied because of its crucial role in the photocatalytic process. The contribution of this photo-generated radical was studied by adding potassium dichromate ($K_2Cr_2O_7$) to the reaction medium. The results of adding are shown in Figure 6d. The presence of potassium dichromate decreased the reaction rate of RG12 discoloration, where a reduction of 32.7% in discoloration efficiency was recorded in the presence of 1.1 mmol/L $K_2Cr_2O_7$. An inversely proportional relationship between the added quantity of $K_2Cr_2O_7$ and the discoloration efficiency of RG12 was also noticed. This can be explained by the positive role played by $O_2{}^{\bullet-}$ in the elimination of RG12 textile dye.

3.4. Effect of the Addition of Oxidizing Agents: Cases of H_2O_2 and $K_2S_2O_8$

In order to enhance the photocatalytic reaction, several intensification ways can be tested. The effect of the adding oxidants such as hydrogen peroxide (H_2O_2) and potassium peroxydisulfate ($K_2S_2O_8$) in the reaction medium (mimicking real textile industry effluents) on the photocatalytic reaction kinetics was investigated. Figure 7a shows that adding hydrogen peroxide (H_2O_2) strongly improved the reaction rate, where 47.4% and 100% discoloration were, respectively, achieved after 30 and 300 min in the presence of 4.6 mmol/L H_2O_2. However, only 48% of discoloration efficiency was noticed without any addition after 300 min under light. The strong enhancement caused by the presence of H_2O_2 can be explained by (i) the participation of H_2O_2 in the generation of $^{\bullet}OH$ from one hand [44] and (ii) the reduction of the recombination rate as shown below in Equations (2)–(4) from another hand [45,46]. It is worth to mention at this level that the $^{\bullet}OH$ is 1.5 times more oxidant than H_2O_2 and 1.4 times more than ozone [47,48].

$$e^-{}_{cb} + H_2O_2 \rightarrow 2HO^{\bullet} \tag{2}$$

$$O_2{}^{\bullet-} + H_2O_2 \rightarrow {}^{\bullet}OH + OH^- + O_2 \tag{3}$$

$$hv + H_2O_2 \rightarrow 2({}^{\bullet}OH) \tag{4}$$

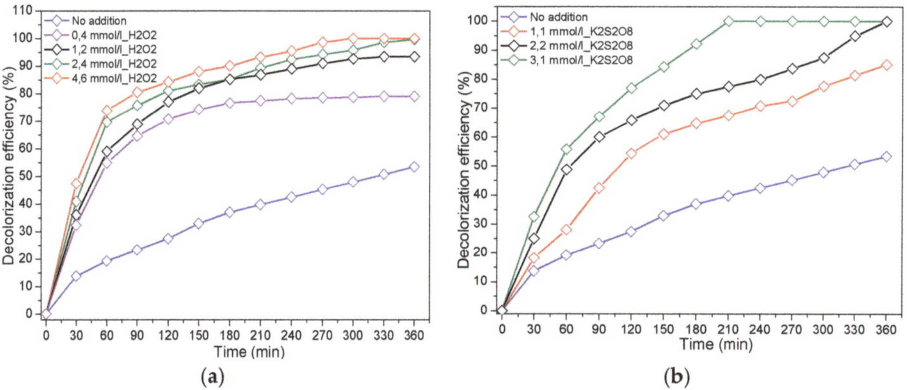

Figure 7. Photocatalytic degradation of RG12 in the presence of (**a**) hydrogen peroxide (H_2O_2) and (**b**) potassium peroxydisulfate ($K_2S_2O_8$). (Initial RG12 concentration = 4 mg/L; sample, HiPIMS at 40 A; and sputtering time, 100 s.)

The effect of persulfate ions on the photocatalytic degradation of RG12 was investigated by varying its concentration from 1.1 to 3.1 mmol/L. Figure 7b shows that the presence of potassium peroxydisulfate ($K_2S_2O_8$) drastically increased the reaction rate of RG12 discoloration, where 67.5% and a full discoloration (100%) were achieved within 210 min with 1.1 and 3.1 mmol/L $S_2O_8{}^{2-}$, respectively. This can be due to the electron-scavenging properties of $S_2O_8{}^-$ [49], which led to the formation of sulfate radical anions ($SO_4{}^{\bullet-}$) with a high oxidizing power ($E° = 2.6$ eV) by $S_2O_8{}^{2-}$ reduction according to the Equations (5) and (6) [50].

$$S_2O_8{}^{2-} + e^-{}_{cb} \rightarrow SO_4{}^{\bullet-} + SO_4{}^{2-} \tag{5}$$

$$SO_4{}^{\bullet-} \rightarrow SO_4{}^{2-} + h_{vb}{}^+ \tag{6}$$

The effect of $SO_4{}^{\bullet-}$ in the enhancement of RG12 discoloration can be rationalized in (i) the prevention of electron/hole recombination rate resulted from the interaction with conduction band electrons leaving behind positives holes [51]; (ii) the abstraction of a hydrogen atom from saturated carbon [44]; (iii) the generation of hydroxyl radicals by interaction with H_2O molecules according to Equation (7) [52]; and (iv) the possible reaction of sulfate radical anions and the dye molecules by direct attack [53].

$$SO_4^{\bullet-} + H_2O \rightarrow SO_4^{2-} + {}^{\bullet}OH + H^+ \tag{7}$$

At this stage, it is worth to mention that copper oxide can be easily "sulfated". This may affect the catalytic performance of the catalyst.

3.5. Effect of Different Water Matrices on RG12 Degradation

In order to approach a real case of textile effluent, the effect of mineral ions' and salts' presence on the photocatalytic reaction rate was investigated by testing both Cu and NaCl as inorganic ions. Cu and NaCl were tested separately, and then combined together. It is readily seen from Figure 8 that the presence of Cu has a negative effect on the discoloration efficiency which can be explained as follows: (i) the possible complexation of Cu and organic species or some intermediate by-products [54]; (ii) according to the work of Kumawar et al. [55], there is an optimum concentration when adsorbed metal ions change the behavior from enhancer to inhibitor because of changing surface charge of the photocatalyst or the target pollutant. Similar results on Cu-inhibitory effect was found by Tercero Espinoza et al. [56]; and (iii) the presence of transition metal ions can alter the electron-transfer pathway, decrease the reduction of oxygen, and even suppress the degradation [57].

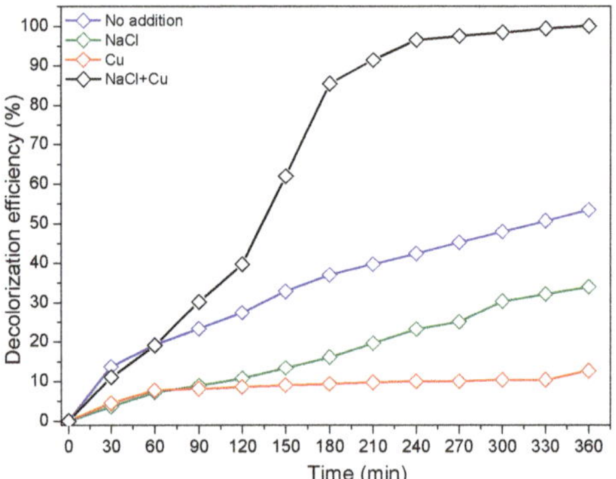

Figure 8. Photocatalytic degradation of RG12 in the presence NaCl and Cu. (Initial RG12 concentration = 4 mg/L; sample, HiPIMS at 40 A; and sputtering time, 100 s).

The addition of NaCl as chloride ions source (Cl^-) to the dye solution in our experiment led to a reduction of discoloration efficiency by about 20% after 360 min under LEDs light (420 nm).

The effect of anions in the photocatalytic medium has been reported to have an inhibitory characters due to (1) the competitive effect between the anions and the target pollutant toward the active sites available on the surface of the photocatalyst [58]; (2) the anion radicals' scavenging properties such as holes and hydroxyl radicals, which led to the formation of ionic radicals such as $Cl^{\bullet}$, $NO_3^{\bullet}$, and $HCO_3^{\bullet-}$ less reactive than ${}^{\bullet}OH$ [20]; and (3) the h^+-scavenging proprieties of chloride ions according to Equation (8) [59].

$$Cl^- + h^+{}_{vb} \rightarrow Cl^{\bullet-} + H^+ \tag{8}$$

The simultaneous addition of Cu and NaCl (both at the same time) led to a drastic enhancement in the discoloration rate of RG12, where 98% of RG12 discoloration was observed within 360 min of illumination, as shown in Figure 8.

RG12 removal can be influenced by the multitude of species present in the real wastewater effluent. Mathematical simulation of the degradation of RG12 was seen to fit a second-order model as described by the equation (see supplementary Table S1):

$$\frac{d[\text{RG12}]}{dt} = -k[\text{RG12}][\text{AS}]$$

where [AS] is the steady-state concentration of actives species (radical hydroxyl, sulfate ion radical, reactive oxygen anion superoxide, etc.), [RG12] is the concentration of RG12 in water, k is the second-order rate constant and t is the reaction time.

Taking into account the instantaneous concentrations of the photo-generated ROS, the kinetics of the degradation of RG12 in water can be described according to the pseudo-first-order equation as given below:

$$[\text{RG12}](t) = [\text{RG12}]_0 \exp^{-k_{app}\cdot t}$$

where k_{app}, is the pseudo-first-order apparent rate constant (min^{-1}) and it was obtained by linear regression of $\ln(C_t/C_0)$ vs. time t.

Figure 9 summarizes the reaction rate constants k_{app} of RG12 degradation with and without the addition of various chemicals (scavengers, oxidants, or inorganic pollutants). The constants varied from very low values of 0.00051 min^{-1} in the presence of Cu ions to very high values (0.0143 min^{-1}) in the presence of H_2O_2. In the presence of H_2O_2, there was an enhancement of 7 times of the RG12 discoloration rate followed by a 6 times' enhancement with $K_2S_2O_8$. It is readily seen that the constant rate of reaction in the presence of methanol (0.004 min^{-1}) is higher than with the addition of iso-propanol (0.002 min^{-1}).

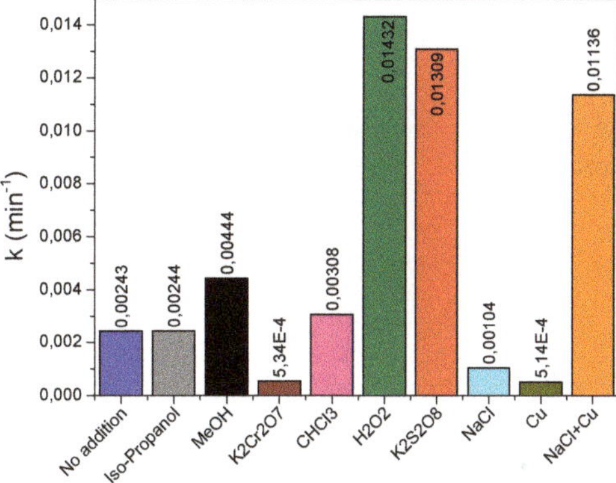

Figure 9. Reaction rate constant of RG12 photocatalytic degradation in the presence of different chemical products.

In the aim to evaluate the stability and the reusability of the synthesized material supported on PES used in this study, successive photocatalytic cycles were applied for the degradation of RG12 (4 mg/L in each experiment) with the same catalyst. As it can be seen in Figure 10, the photocatalyst has a good stability under LEDs illumination approved by the slight decrease in discoloration efficiency (about 7%) after 21 runs.

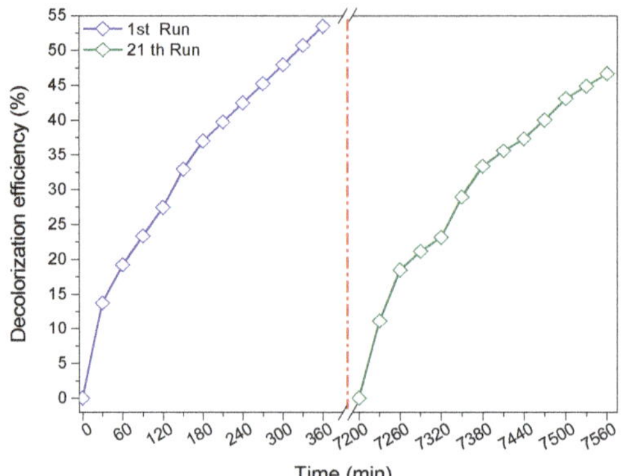

Figure 10. Recycling runs of the catalyst for the photocatalytic degradation of RG12 up to 21 runs (initial concentrations: 4 mg/L; catalyst dose, 1 sheet; current intensity, 40 A; and sputtering time, 100 s).

Table 2 shows the released Cu and Ti ions from the fabrics after 1, 5, 10, and 21 washing cycles. The results presented in Table 2 are cumulative quantification of the released ions during all the cycles. It is readily seen from Table 2 that the Ti ions are found at very low quantities during the first 5 cycles and then reduced drastically to almost zero ppb at the end of the recycling experiment. Table 2 also shows that the Cu ions' release slows down at the end of the recycling (after the 10th cycle) to stabilize at about 2–3 ppb/cycle. These results were slightly similar to previous reports using HiPIMS deposition of Cu_xO or TiO_2/Cu deposited on polyester for photocatalytic applications [15,31,60]. The cumulative ions' release is still far below the threshold fixed by regulatory bodies for Cu and/or Ti ions.

Table 2. Inductively coupled plasma mass spectrometry (ICP-MS) quantification of ions' release during the recycling experiment.

	Cu Ions	**Ti Ions**
Cycle 1	17	6
Cycle 5	29	9
Cycle 10	32	10
Cycle 21	34	10

4. Conclusions

Cu_xO thin films deposited using HiPIMS on polyester under different sputtering energies were successfully synthesized. The sputtering mode and the applied current intensity were optimized. The photocatalytic performance of the photocatalyst was evaluated for the degradation of RG12 under visible light LEDs irradiation with different initial dye concentrations and surface area to volume ratios. The effect of different concentrations of various chemicals (scavengers, oxidants, or inorganic ions) on the photocatalytic process was also studied mimicking a real textile effluent containing a large variety of compounds. The monitoring of ROS showed that the superoxide ions played the main role in RG12 degradation contrary to hydroxyl radicals and the photo-generated holes. The presence of H_2O_2 or $K_2S_2O_8$ presented a high oxidizing power, enhancing the photocatalytic activity of the sputtered catalyst. The presence of Cu as mineral pollution and NaCl as salt decreased the reaction rate, but the addition of both of them (Cu and NaCl salts) improved the removal efficiency of the studied pollutant. The recycling of the catalyst showed a high stability of the catalyst up to 21 RG12 discoloration cycles. ICP-MS showed stable ions' release after the 5th cycle for both ions. This allows potential industrial applications of the reported HiPIMS coatings in the future.

Supplementary Materials: The following are available online at http://www.mdpi.com/1996-1944/12/3/412/s1, Table S1: Second order rate constants of different scavengers with radical species.

Author Contributions: Conceptualization: S.R and A.A.A.; Methodology: A.A.A., A.A. and S.R.; Software and validation: H.Z., N.K. and H.D.; Investigation and Experimental: S.R., H.Z. and A.A.A.; Resources: S.R., A.A. and A.A.A.; Writing—original draft preparation: H.Z. and A.A.A.; Writing—review and editing A.A. and S.R.; Supervision: A.A., A.A.A., H.D. and N.K.

Acknowledgments: S.R. acknowledges the support of EPFL-IPhys (especially, Prof. László Forró) and M. Bensimon. Authors would like to thank ENSCR support.

Conflicts of Interest: The authors declare no conflict of interest.

References

1. Houas, A.; Lachheb, H.; Ksibi, M.; Elaloui, E.; Guillard, C.; Herrmann, J.-M. Photocatalytic degradation pathway of methylene blue in water. *App. Cat. B Environ.* **2001**, *31*, 145–157. [CrossRef]

2. Grzechulska, J.; Morawski, A.W. Photocatalytic Decomposition of Azo-Dye Acid Black 1 in Water over Modified Titanium Dioxide. *Appl. Catal. B Environ.* **2002**, *36*, 45–51. [CrossRef]

3. Radhika, N.P.; Selvin, R.; Kakkar, R.; Umar, A. Recent advances in nano-photocatalysts for organic synthesis. *Arab. J. Chem.* **2016**. [CrossRef]

4. Šegota, S.; Ćurković, L.; Ljubas, D.; Houra, I.F.; Tomašić, N. Synthesis, characterization and photocatalytic properties of sol–gel TiO$_2$ films. *Ceram. Int.* **2011**, *37*, 1153–1160. [CrossRef]

5. Zulfiqar, M.; Chowdhury, S.; Omar, A.A. Hydrothermal synthesis of multiwalled TiO$_2$ nanotubes and its photocatalytic activities for Orange II removal. *Sep. Sci. Technol.* **2018**, *53*, 1412–1422. [CrossRef]

6. Cheng, H.-H.; Chen, S.-S.; Yang, S.-Y.; Liu, H.-M.; Lin, K.-S. Sol-Gel Hydrothermal Synthesis and Visible Light Photocatalytic Degradation Performance of Fe/N Codoped TiO$_2$ Catalysts. *Materials* **2018**, *11*, 939. [CrossRef]

7. Yamashita, H.; Harada, M.; Misaka, J.; Takeuchi, M.; Ikeue, K.; Anpo, M. Degradation of propanol diluted in water under visible light irradiation using metal ion-implanted titanium dioxide photocatalysts. *J. Photochem. Photobiol. A* **2002**, *148*, 257–261. [CrossRef]

8. Nagamine, S.; Sugioka, A.; Iwamoto, H.; Konishi, Y. Forma tion of TiO$_2$ hollow microparticles by spraying water drop lets into an organic solution of titanium tetraisopropoxide (TTIP)—Effects of TTIP concentration and TTIP protecting additives. *Powder Technol.* **2008**, *186*, 168–175. [CrossRef]

9. Hosseinnia, A.; Keyanpour-Rad, M.; Pazouki, M. Photo-catalytic Degradation of Organic Dyes with Different Chromophores by Synthesized Nanosize TiO$_2$ Particles. *World Appl. Sci. J.* **2010**, *8*, 1327–1332.

10. Shinde, D.R.; Tambade, P.S.; Chaskar, M.G.; Gadave, K.M. Photocatalytic degradation of dyes in water by analytical reagent grades ZnO, TiO$_2$ and SnO$_2$: A comparative study, Drink. *Water Eng. Sci.* **2017**, *10*, 109–117. [CrossRef]

11. Sacco, O.; Stoller, M.; Vaiano, V.; Ciambelli, P.; Chianese, A.; Sannino, D. Photocatalytic Degradation of Organic Dyes under Visible Light on N-Doped TiO$_2$ Photocatalysts. *Int. J. Photoenergy* **2012**, *2012*, 626759. [CrossRef]

12. Habba, Y.G.; Capochichi-Gnambodoe, M.; Leprince-Wang, Y. Enhanced Photocatalytic Activity of Iron-Doped ZnO Nanowires for Water Purification. *Appl. Sci.* **2017**, *7*, 1185. [CrossRef]

13. Gionco, C.; Fabbri, D.; Calza, P.; Paganini, M.C. Synthesis, Characterization and Photocatalytic Tests of N-Doped Zinc Oxide: A New Interesting Photocatalyst. *J. Nanomater.* **2016**, *2016*, 4129864. [CrossRef]

14. Mayoufi, A.; Nsib, M.F.; Ahmed, O.; Houas, A. Synthesis, characterization and photocatalytic performance of W, N, S-tri-doped TiO$_2$ under visible light irradiation. *C. R. Chim.* **2015**, *18*, 875–882. [CrossRef]

15. Baghriche, O.; Rtimi, S.; Pulgarin, C.; Kiwi, J. Polystyrene CuO/Cu$_2$O uniform films inducing MB-degradation under sunlight. *Catal. Today* **2017**, *284*, 77–83. [CrossRef]

16. Zeghioud, H.; Assadi, A.A.; Khellaf, N.; Djelal, H.; Amrane, A.; Rtimi, S. Reactive species monitoring and their contribution for removal of textile effluent with photocatalysis under UV and visible lights: Dynamics and mechanism. *J. Photochem. Photobiol. A Chem.* **2018**, *365*, 94–102. [CrossRef]

17. Aliah, H.; Aji, M.P.; Sustini, E.; Budiman, M.; Abdullah, M. TiO$_2$ Nanoparticles-Coated Polypropylene Copolymer as Photocatalyst on Methylene Blue Photodegradation under Solar Exposure. *Am. J. Environ. Sci.* **2012**, *8*, 280–290.

18. Zeghioud, H.; Khellaf, N.; Amrane, A.; Djelal, H.; Elfalleh, W.; Assadi, A.A.; Rtimi, S. Photocatalytic performance of TiO$_2$ impregnated polyester for the degradation of Reactive Green 12: Implications of the surface pretreatment and the microstructure. *J. Photochem. Photobiol. A Chem.* **2017**, *346*, 493–501. [CrossRef]

19. Rtimi, S. Indoor Light Enhanced Photocatalytic Ultra-Thin Films on Flexible Non-Heat Resistant Substrates Reducing Bacterial Infection Risks. *Catalysts* **2017**, *7*, 57. [CrossRef]

20. Zeghioud, H.; Khellaf, N.; Djelal, H.; Amrane, A.; Bouhelassa, M. Photocatalytic reactors dedicated to the degradation of hazardous organic pollutants: Kinetics, mechanistic aspects and Design—A review. *Chem. Eng. Commun.* **2016**, *203*, 1415–1431. [CrossRef]

21. Hoffmann, M.R.; Martin, S.T.; Choi, W.; Bahnemann, D.W. Environmental applications of semiconductor photocatalysis. *Chem. Rev.* **1995**, *95*, 69–96. [CrossRef]

22. Biswas, M.R.D.; Ali, A.; Cho, K.Y.; Oh, W.-C. Novel synthesis of WSe$_2$-Graphene-TiO$_2$ ternary nanocomposite via ultrasonic technics for high photocatalytic reduction of CO$_2$ into CH$_3$OH. *Ultrason. Sonochem.* **2018**, *42*, 738–746. [CrossRef] [PubMed]

23. Fujishima, A.; Rao, T.N.; Tryk, D.A. Titanium dioxide photocatalysis. *J. Photochem. Photobiol. C Photochem. Rev.* **2000**, *1*, 1–21. [CrossRef]

24. Rtimi, S.; Dionysiou, D.D.; Pillai, S.C.; Kiwi, J. Advances in bacterial inactivation by Ag, Cu, Cu-Ag coated surfaces & medical devices. *Appl. Catal. B Environ.* **2019**, *240*, 291–318.

25. Rtimi, S.; Pulgarin, C.; Sanjines, R.; Kiwi, J. Kinetics and mechanism for transparent polyethylene-TiO$_2$ films mediated self-cleaning leading to MB dye discoloration under sunlight irradiation. *Appl. Catal. B Environ.* **2015**, *162*, 236–244. [CrossRef]

26. Tan, X.-Q.; Liu, J.-Y.; Niu, J.-R.; Liu, J.-Y.; Tian, J.-Y. Recent progress in magnetron sputtering technology used on fabrics. *Materials* **2018**, *11*, 1953. [CrossRef] [PubMed]

27. Uyguner-Demirel, C.S.; Birben, C.N.; Bekbolet, M. A comprehensive review on the use of second generation TiO$_2$ photocatalysts: Microorganism inactivation. *Chemosphere* **2018**, *211*, 420–448. [CrossRef] [PubMed]

28. Ballo, M.K.S.; Rtimi, S.; Kiwi, J.; Pulgarin, C.; Entenza, J.M.; Bizzizi, A. Fungicidal activity of copper-sputtered flexible surfaces under dark and actinic light against azole-resistant *Candida albicans* and *Candida glabrata*. *J. Photochem. Photobiol. B Biol.* **2017**, *174*, 229–234. [CrossRef]

29. Wu, H.; Zhang, X.; Geng, Z.; Yin, Y.; Hang, R.; Huang, X.; Yao, X.; Tang, B. Preparation, antibacterial effects and corrosion resistant of porous Cu–TiO$_2$ coatings. *Appl. Surf. Sci.* **2014**, *308*, 43–49. [CrossRef]

30. Sangchay, W.; Sikong, L.; Kooptarnond, K. Photocatalytic and Self-Cleaning Properties of TiO$_2$-Cu Thin Films on Glass Substrate. *Appl. Mech. Mater.* **2012**, *152–154*, 409–413. [CrossRef]

31. Rtimi, S.; Baghriche, O.; Pulgarin, C.; Lavanchy, J.; Kiwi, J. Growth of TiO$_2$/Cu films by HiPIMS for accelerated bacterial loss of viability. *Surf. Coat. Technol.* **2013**, *232*, 804–813. [CrossRef]

32. Sadeghi-Kiakhani, M.; Khamseh, S.; Rafie, A.; Tekieh, S.M.F.; Zarrintaj, P.; Saeb, M.R. Thermally stable antibacterial wool fabrics surface-decorated by TiON and TiON/Cu thin films. *Surf. Innov.* **2018**, *6*, 258–265. [CrossRef]

33. *Active Coatings for Smart Textiles, Chapter 3: Environmentally Mild Self-Cleaning Processes on Textile Surfaces under Daylight Irradiation: Critical Issues*; Kiwi, J.; Rtimi, S. (Eds.) Elsevier BV.: Amsterdam, The Netherlands, 2016; pp. 35–54.

34. Rtimi, S.; Sanjines, R.; Pulgarin, C.; Kiwi, J. Microstructure of Cu−Ag Uniform Nanoparticulate Films on Polyurethane 3D Catheters: Surface Properties. *ACS Appl. Mater. Interfaces* **2016**, *8*, 56–63. [CrossRef] [PubMed]

35. Docampo, P.; Guldin, S.; Steiner, U.; Snaith, H.J. Charge Transport Limitations in Self-Assembled TiO$_2$ Photo-anodes for Dye-Sensitized Solar Cells. *J. Phys. Chem. Lett.* **2013**, *4*, 698–703. [CrossRef] [PubMed]

36. Mamba, G.; Pulgarin, C.; Kiwi, J.; Bensimon, M.; Rtimi, S. Synchronic coupling of Cu$_2$O(p)/CuO(n) semiconductors leading to Norfloxacin degradation under visible light: Kinetics, mechanism and film surface properties. *J. Catal.* **2017**, *353*, 133–140. [CrossRef]

37. Mao, P.; Jiang, J.; Pan, Y.; Duanmu, C.; Chen, S.; Yang, Y.; Zhang, S.; Chen, Y. Enhanced Uptake of Iodide from Solutions by Hollow Cu-Based Adsorbents. *Materials* **2018**, *11*, 769. [CrossRef]

38. Huang, M.; Xu, C.; Wu, Z.; Huang, Y.; Lin, J.; Wu, J. Photocatalytic discolorization of methyl orange solution by Pt modified TiO_2 loaded on natural zeolite. *Dyes Pigm.* **2008**, *77*, 327–334. [CrossRef]

39. Rong, S.; Sun, Y. Degradation of TAIC by water falling film dielectric barrier discharge—Influence of radical scavengers. *J. Hazard. Mater.* **2015**, *287*, 317–324. [CrossRef]

40. Giraldo, A.L.; Penuela, G.A.; Torres-Palma, R.A.; Pino, N.J.; Palominos, R.A.; Mansilla, H.D. Degradation of the antibiotic oxolinic acid by photocatalysis with TiO_2 in suspension. *Water Res.* **2010**, *44*, 5158–5167. [CrossRef]

41. Chen, Y.; Yang, S.; Wang, K.; Lou, L. Role of primary active species and TiO_2 surface characteristic in UV-illuminated photodegradation of acid orange 7. *J. Photochem. Photobiol. A Chem.* **2005**, *172*, 47–54. [CrossRef]

42. di Valentin, C.; Fittipaldi, D. Hole Scavenging by Organic Adsorbates on the TiO_2 Surface: A DFT Model Study. *J. Phys. Chem. Lett.* **2013**, *4*, 1901–1906. [CrossRef] [PubMed]

43. Shen, M.; Henderson, M.A. Identification of the Active Species in Photochemical Hole Scavenging Reactions of Methanol on TiO_2. *J. Phys. Chem. Lett.* **2011**, *2*, 2707–2710. [CrossRef]

44. Pare, B.; Jonnalagadda, S.B.; Tomar, H.; Singh, P.; Bhagwat, V.W. ZnO assisted photocatalytic degradation of acridine orange in aqueous solution using visible irradiation. *Desalination* **2008**, *232*, 80–90. [CrossRef]

45. Gupta, V.K.; Jain, R.; Agarwal, S.; Nayak, A. Photodegradation of hazardous dye quinoline yellow catalyzed by TiO_2, Meenakshi Shrivastava. *J. Colloid Interface Sci.* **2012**, *366*, 135–140. [CrossRef] [PubMed]

46. Rodríguez, E.M.; Márquez, G.; Tena, M.; Álvarez, P.M.; Beltrán, F.J. Determination of main species involved in the first steps of TiO_2 photocatalytic degradation of organics with the use of scavengers: The case of ofloxacin. *Appl. Catal. B* **2015**, *178*, 44–53. [CrossRef]

47. Al-Ekabi, H.A.; Serpone, N. Kinetics studies in heterogeneous photocatalysis. I. Photocatalytic degradation of chlorinated phenols in aerated aqueous solutions over titania supported on a glass matrix. *J. Phys. Chem.* **1988**, *92*, 5726–5731. [CrossRef]

48. Fujihira, M.; Satoh, Y.; Osa, T. Heterogeneous photocatalytic reaction on semiconductor materials. III.effect of pH and Cu^+ ions on the photo-fenton type reaction. *Bull. Chem. Soc. Jpn.* **1982**, *55*, 666–671. [CrossRef]

49. Das, D.P.; Baliarsingh, N.; Parida, K.M. Photocatalytic decolorisation of Methylene Blue (MB) over titania pillared zirconium phosphate (ZrP) and titanium phosphate (TiP) under solar radiation. *J. Mol. Catal. A Chem.* **2007**, *261*, 241–261. [CrossRef]

50. Sharma, A.; Dutta, R.K. Studies on drastic improvement of photocatalytic degradation of acid orange -74 dye by TPPO capped CuO nanoparticles in tandem with suitable electron capturing agents. *RSC Adv.* **2015**, *5*, 43815–43823. [CrossRef]

51. Subramonian, W.; Wu, T. Effect of Enhancers and Inhibitors on Photocatalytic Sunlight Treatment of Methylene Blue. *Water Air Soil Pollut.* **2014**, *225*, 1–15. [CrossRef]

52. Rupa, A.V.; Manikandan, D.; Divakar, D.; Sivakumar, T. Effect of deposition of Ag on TiO_2 nanoparticles on the photodegradation of Reactive Yellow-17. *J. Hazard. Mater.* **2007**, *147*, 906–913. [CrossRef] [PubMed]

53. Neppolian, B.; Choi, H.C.; Sakthivel, S.; Arabindoo, B.; Murugesan, V. Solar light induced and TiO_2 assisted degradation of textile dye Reactive Blue 4. *Chemosphere* **2002**, *46*, 1173–1181. [CrossRef]

54. Chaudharya, A.J.; Hassana, M.U.; Grimes, S.M. Simultaneous recovery of metals and degradation of organic species: Copper and 2,4,5-trichlorophenoxyacetic acid (2,4,5-T). *J. Hazard. Mater.* **2009**, *165*, 825–831. [CrossRef] [PubMed]

55. Kumawat, R.; Bhati, I.; Ameta, R. Role of some metal ions in photocatalytic degradation of Rose Bengal dye. *Indian J. Chem. Technol.* **2012**, *19*, 191–194.

56. Espinoza, L.A.T.; Haseborg, E.T.; Weber, M.; Karle, E.; Peschke, R.; Frimmel, F.H. Effect of selected metal ions on the photocatalytic degradation of bog lake water natural organic matter. *Water Res.* **2011**, *45*, 1039–1048. [CrossRef]

57. Chen, C.; Li, X.; Ma, W.; Zhao, J.; Hidaka, H.; Serpone, N. Effect of Transition Metal Ions on the TiO_2-Assisted Photodegradation of Dyes under Visible Irradiation: A Probe for the Interfacial Electron Transfer Process and Reaction Mechanism. *J. Phys. Chem. B* **2002**, *106*, 318–324. [CrossRef]

58. Bahnemann, D.; Robertson, P. (Eds.) Environmental Photochemistry Part III. In *The Handbook of Environmental Chemistry D, Surface-Modified Photocatalysts*; Springer: Berlin, Germany, 2015; Volume 35, pp. 23–44.

59. Bockelmann, D.; Lindner, M.; Bahnemann, D. *Fine Particles: Science and Technology*; Kluwer Academic Publishers: Amsterdam, The Netherlands, 1996.
60. Rtimi, S.; Konstantinidis, S.; Britun, N.; Bensimon, M.; Khmel, I.; Nadtochenko, V. Extracellular bacterial inactivation proceeding without Cu-ion release: Drastic effects of the applied plasma energy on the performance of the Cu-polyester (PES) samples. *Appl. Catal. B Environ.* **2018**, *239*, 245–253. [CrossRef]

Article

Influence of Titanium Dioxide Nanoparticles on the Sulfate Attack upon Ordinary Portland Cement and Slag-Blended Mortars

Atta-ur-Rehman, Abdul Qudoos, Hong Gi Kim and Jae-Suk Ryou *

Department of Civil and Environmental Engineering, Hanyang University, 222, Wangsimni-ro, Seongdong-gu, Seoul 04763, Korea; attabrcian@gmail.com (A.R.); qudoos.engnr@gmail.com (A.Q.); dmkg1404@naver.com (H.G.K.)
* Correspondence: jsryou@hanyang.ac.kr; Tel.: +82-2-2220-4323

Received: 30 January 2018; Accepted: 27 February 2018; Published: 28 February 2018

Abstract: In this study, the effects of titanium dioxide (TiO_2) nanoparticles on the sulfate attack resistance of ordinary Portland cement (OPC) and slag-blended mortars were investigated. OPC and slag-blended mortars (OPC:Slag = 50:50) were made with water to binder ratio of 0.4 and a binder to sand ratio of 1:3. TiO_2 was added as an admixture as 0%, 3%, 6%, 9% and 12% of the binder weight. Mortar specimens were exposed to an accelerated sulfate attack environment. Expansion, changes in mass and surface microhardness were measured. Scanning Electron Microscopy (SEM), Energy Dispersive Spectroscopy (EDS), X-ray Diffraction (XRD), Thermogravimetry Analysis (TGA) and Differential Scanning Calorimetry (DSC) tests were conducted. The formation of ettringite and gypsum crystals after the sulfate attack were detected. Both these products had caused crystallization pressure in the microstructure of mortars and deteriorated the mortars. Our results show that the addition of nano-TiO_2 accelerated expansion, variation in mass, loss of surface microhardness and widened cracks in OPC and slag-blended mortars. Nano-TiO_2 containing slag-blended mortars were more resistant to sulfate attack than nano-TiO_2 containing OPC mortars. Because nano-TiO_2 reduced the size of coarse pores, so it increased crystallization pressure due to the formation of ettringite and gypsum thus led to more damage under sulfate attack.

Keywords: titanium dioxide; nanoparticles; photocatalysis; sulfate attack; mortar; cement; blast furnace slag; expansion; deterioration; microcracks

---

1. Introduction

1.1. Background

Titanium dioxide (TiO_2) is a white and inorganic material. It is used as an effective photocatalyst and can be activated by light radiation to degrade organic and inorganic pollutants present in the water and air through oxidation-reduction process. In recent decades, TiO_2 has been added to concrete during mixing to yield a product with self-cleaning and air purifying properties [1]. TiO_2 can clean the surface of concrete by degrading certain pollutants deposited on its surface in the presence of sunlight and can also convert harmful gases like nitrous oxides and volatile organic compounds (VOCs) to less harmful products [2,3]. Concrete with the addition of TiO_2 has been referred to in the literature as self-cleaning, air purifying, or photocatalytic concrete. Photocatalytic concrete has been effectively used in the pavement blocks, tunnels, sidewalks and external walls of buildings for air purifying purposes and to maintain the aesthetic properties of buildings [4,5]. Additionally, ground granulated blast furnace slag (GGBFS) is an industrial byproduct with a white color. Use of slag in cementitious materials is environmentally beneficial as it helps in reducing carbon dioxide emissions and energy use due to cement production [6]. It can be used in concentrations up to 30% to replace white Portland cement for architectural effects in white concrete structures [7]. The aesthetic appeal of such structures can be preserved by adding nano-TiO_2 [4,8]. Simultaneous use of TiO_2 and slag in cementitious materials can inherit them with beneficial characteristics of both slag and TiO_2.

Sulfate attack is one of the most aggressive environmental conditions to address for concrete. Sulfate ions present in the soil, run-off water, seawater, groundwater and sewer lines can move to the interior of the concrete through pores and react with unhydrated and hydrated alumina phases, portlandite (CH), calcium silicate hydrate (CSH) to produce ettringite and gypsum. Incoming SO_4^{2-} ions react with calcium alumino monosulfate phases and produce ettringite ($3CaO \cdot Al_2O_3 \cdot 3CaSO_4 \cdot 32H_2O$) according to Equation (1) [9]:

$$SO_4^{2-} + 4CaO \cdot Al_2O_3 \cdot SO_3 \cdot 12H_2O + 2Ca^{2+} + 20H_2O \rightarrow 3CaO \cdot Al_2O_3 \cdot 3CaSO_4 \cdot 32H_2O \tag{1}$$

Sodium sulfate reacts with portlandite to produce gypsum and sodium hydroxide according to Equation (2):

$$Na_2SO_4 + Ca(OH)_2 \rightarrow CaSO_4 \cdot 2H_2O + NaOH \tag{2}$$

Gypsum reacts with hydrated products such as calcium aluminates, calcium sulfoaluminate (monosulfate- C_4ASH_{12-18}) or tricalcium aluminate, (unhydrated phase in cement clinker) to produce ettringite, as given in the Equations (3)–(5):

$$4CaO \cdot Al_2O_3 \cdot 13H_2O + 3(CaSO_4 \cdot 2H_2O) + 13H_2O \rightarrow 3CaO \cdot Al_2O_3 \cdot 3CaSO_4 \cdot 32H_2O + Ca(OH)_2 \tag{3}$$

$$4CaO \cdot Al_2O_3 \cdot SO_3 \cdot 12H_2O + 2(CaSO_4 \cdot 2H_2O) + (10 - 16)\,H_2O \rightarrow 3CaO \cdot Al_2O_3 \cdot 3CaSO_4 \cdot 32H_2O \tag{4}$$

$$3CaO \cdot Al_2O_3 + 3(CaSO_4 \cdot 2H_2O) + 26H_2O \rightarrow 3CaO \cdot Al_2O_3 \cdot 3CaSO_4 \cdot 32H_2O \tag{5}$$

Ettringite is an expansive product and causes expansion in the hardened cement paste. Expansion due to gypsum is disputed [10], researchers agree that gypsum softens mortar and reduces its strength [11]. If the expansive stresses from the formation of new products exceed the tensile strength of the hardened concrete, then microcracks are formed. These microcracks allow greater transport of sulfate ions from the external environment to the interior of the concrete and accelerate the sulfate attack. Sulfate attack increases porosity, reduces strength, softens concrete, changes mass, causes expansion, cracking and spalling. The severity and nature of these defects are dependent on factors such as cement type and composition, water/binder ratio, presence and amount of supplementary cementitious materials, porosity, permeability, ambient temperature, the concentration of sulfate ions and types of cations, such as sodium or magnesium. More details relating to the mechanism of sulfate attack can be found in the previous reviews [9,11–13].

1.2. Research Significance

During service life, photocatalytic concrete is in contact with soil or water; as both soil and water are sources of sulfate attack, there is a high possibility of such an attack. The addition of nano-TiO_2 alters the mechanical properties and the microstructure of the concrete [14]; therefore, it is necessary to investigate sulfate attack on nano-TiO_2 containing concrete to design efficient photocatalytic structures and avoid premature failure. Research into sulfate attack on photocatalytic mortars is sparse. In this study, we investigate the effects of nano-TiO_2 on the sulfate attack resistance of pure Portland cement mortars and slag-blended mortars at two exposure temperatures. Nano-TiO_2 was added to the mortars at binder weight percentages of 0%, 3%, 6%, 9% and 12%. Other researchers have also used large TiO_2 dosage in their studies [14–18]. Lower dosage of TiO_2 in the cementitious materials may be insufficient to ensure self-cleaning and air purification at longer ages. Therefore, higher dosage of TiO_2 was chosen. The mortars were immersed in 10% sodium sulfate solution at 5 and 25 °C and expansion, changes in mass and loss of surface microhardness were measured. SEM, EDS, XRD, TGA and DSC tests were conducted on deteriorated specimens.

2. Experimental Program

2.1. Materials and Mix Proportion

Titanium dioxide used in this study was produced by Cristal France SAS (Thann, France). According to Cristal Corporate, the content of anatase phase in the product was more than 99%, the size of nanoparticles ranged from 10~60 nm and specific surface area of the powder was 85.6 m^2/g. Ordinary Portland cement and ground granulated blast furnace slag were purchased from a local company (Dongyong Cement, Co., Ltd., Seoul, Korea). The chemical compositions of the cement and slag are given in Table 1.

Table 1. Chemical composition and physical properties of ordinary Portland cement (OPC) and ground granulated blast furnace slag (GGBFS).

Oxide	OPC	GGBFS
CaO (%)	62.27	40.72
SiO_2 (%)	21.32	34.86
Al_2O_3 (%)	5.19	12.54
SO_3 (%)	2.17	1.32
MgO (%)	3.04	7.61
Fe_2O_3 (%)	2.23	0.72
K_2O (%)	0.58	0.67
Na_2O (%)	0.52	0.38
Loss on ignition	1.5	0.43
Blaine fineness (cm^2/g)	3400	4600
Specific Gravity	3.15	2.9

Control and binder/TiO_2 composite mortars were made by adding TiO_2 as 3%, 6%, 9% and 12% of the binder weight. Water/binder ratio was fixed to 0.4, a polycarboxylate ether-based water reducing admixture (0–1.2% of the binder weight) was added to adjust the fluidity of the mortars. Mix proportions are shown in Table 2. The TiO_2 particles were first deagglomerated and dispersed in water through ultrasonication using a sonic probe for at least 45 min. Then water reducing admixture was added in the water and stirred for one more minute. This mixture was slowly added to the sand and binder in the mortar mixer and then mortar specimens were formed.

Table 2. Mix proportion.

Acronym	OPC (g)	Slag (g)	TiO_2 (g)	Sand (g)	Water/Binder Ratio	Water Reducing Admixture (%)
0% OPC	1500	0	0	4500	0.40	0
3% OPC	1500	0	45	4500	0.40	0.25
6% OPC	1500	0	90	4500	0.40	0.6
9% OPC	1500	0	135	4500	0.40	1
12% OPC	1500	0	180	4500	0.40	1.2
0% slag-blended	750	750	0	4500	0.40	0
3% slag-blended	750	750	45	4500	0.40	0.2
6% slag-blended	750	750	90	4500	0.40	0.55
9% slag-blended	750	750	135	4500	0.40	1
12% slag-blended	750	750	180	4500	0.40	1.2

2.2. Test Procedure and Specimen Preparation

ASTM C 1012 [19] is one of the test methods used to evaluate sulfate resistance. According to this method, mortar prisms with dimensions of 25.4 mm × 25.4 mm × 279.4 mm with pins at both ends are prepared and stored in limewater until reaching a strength of 20 MPa. They are then stored in 5% sodium sulfate (Na_2SO_4) solution and any expansion is measured. The ASTM C 1012 process takes several months to a year to complete; many accelerated tests have been proposed using small specimen sizes and higher concentrations of sulfate solution. A summary of these tests can be found in the work of Kim Van Tittelboom et al. [20]. Ferraris et al. [21] developed a new accelerated test technique for measuring the sulfate resistance of mortar. Here, authors used their test method with some modifications. They had immersed 10 mm × 10 mm × 40 mm rectangular prism specimens in 5% Na_2SO_4 solution. In this study, cylindrical mortar specimens with a diameter of 10 mm and length 40 mm were immersed in 10% Na_2SO_4 solution. Because expansion depends on size rather than shape, specimen size is a crucial factor in evaluating sulfate resistance [20,22], although crack patterns might differ in cylindrical and rectangular prisms. To cast the specimens, molds were made in the laboratory using the barrel of syringes with an internal diameter of 10 mm. The barrel of a syringe was cut from the plunger and needle side to a length of 40 mm and was then fixed on a glass plate with epoxy, as shown in Figure 1a. The mortar was cast in molds in four layers and each layer was compacted 20 times with a 2 mm steel rod and the surface was leveled. Twenty specimens were made for each mix proportion, for a total of 200 specimens. Mortar cubes and prisms were casted to measure compressive and flexural strengths After 48 h of casting, the mortar specimens were removed from the molds and placed in saturated limewater for 28 days. Four specimens of each mix proportion were left immersed in saturated limewater while other specimens were selected for immersion in Na_2SO_4 solution. The studs were firmly fixed on the ends of these mortar specimens with epoxy. To allow sulfate ions to attack only from the sides, the epoxy was applied to both end faces of the specimens. Then they were immersed in 10% Na_2SO_4 solution as shown in Figure 1b, which was renewed every other week. The volumetric ratio of the solution to the test specimens was 20:1. Half of the specimens were stored at a temperature of 5 ± 1 °C and a half at 25 ± 1 °C. Both temperatures were maintained constant for the entirety of the study period. Compressive and flexural strength tests were conducted according to ASTM C109 [23] standard and ASTM C348 [24], respectively, at the age of 28 days. Water absorption was measured at the age of 28 days [25].

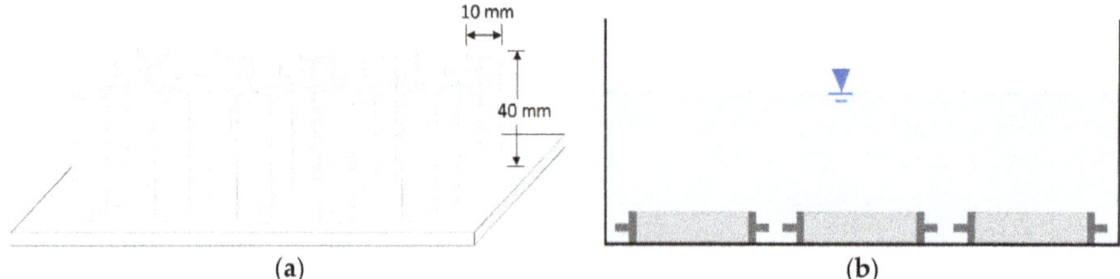

Figure 1. (**a**) A schematic diagram of molds prepared in the laboratory for casting cylindrical mortar specimens; (**b**) Mortar specimens immersed in Na_2SO_4 solution.

2.3. Expansion and Mass Variation

The length and mass of the specimens were measured every seventh day during the immersion period using Equations (6) and (7):

$$\text{Increase in length at (t)} = (L_t - L_i)/L_i \times 100 \tag{6}$$

$$\text{Increase in mass at (t)} = (M_t - M_i)/M_i \times 100 \tag{7}$$

where L is the length, t is the time, i is the initial and M is the mass. The test was stopped after 84 days of immersion.

2.4. Vickers Microhardness

The Vickers microhardness test is a common surface hardness characterization technique [26]. In this method, a static load is applied for a fixed time on the surface of the material using an indenter and the area of indentation is calculated. The Vickers hardness indenter resembles a diamond pyramid in shape. The Vickers hardness value is the ratio of applied load to indentation contact area and is calculated according to Equation (8):

$$HV = 2Psin(\phi 2)/D^2 = 1.854P/D^2,\tag{8}$$

where HV is the Vickers hardness value, P is the load (kgf), D is the mean diagonal of the indentation pyramid (mm) and $\phi = 136°$. To measure the Vickers microhardness, samples with a diameter of 10 mm and depth of 10 mm were obtained by cutting cylindrical specimens with a diamond saw cutter as shown in Figure 2a. Both ends of the obtained samples were successively polished with 400, 800, 1200 and 1500 grit polishing papers and placed in an oven at 55 °C for 24 h to remove the humidity. A load of 0.1 kgf was applied and maintained for 10 s. Microhardness values were measured at 1, 2, 3, 4 and 5 mm from the external surface, as shown in Figure 2b. A minimum of 8 readings were taken at each depth. The test was conducted using a Vickers hardness tester THV-1MD (Capital Instrument, Beijing, China).

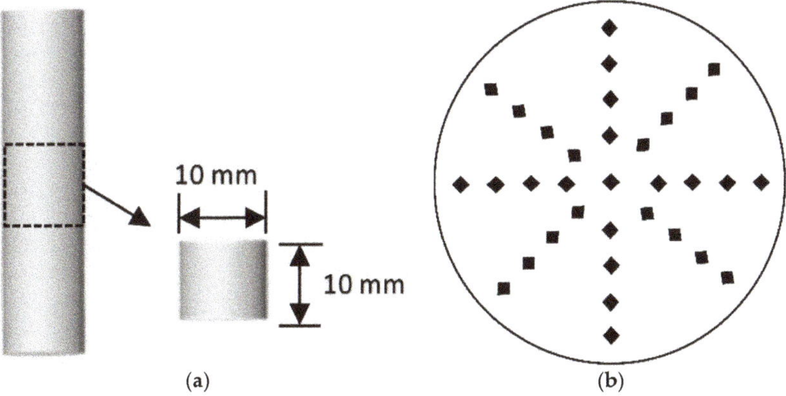

Figure 2. (a) The preparation of the sample used for microhardness; (b) Cross-sectional schematics of the sample used in the Vickers microhardness test.

2.5. Microscopic Observation and Damage Rating

Cracks were observed with a microscope, EGVM-452 M (EG Tech, Seoul, Korea). The samples were observed every four weeks both with the unaided eye and through the microscope to quantify the damage. The specimens were rated on a scale from 0 to 7, as given in Table 3.

2.6. XRD, SEM, EDS, TGA and DSC Analyses

The damaged specimens were analyzed using X-ray diffraction (XRD, Rigaku, Tokyo, Japan), Scanning Electron Microscopy (SEM; Model S-3000 N, Hitachi, Tokyo, Japan) equipped with Energy Dispersive Spectroscopy (EDS) tester after 84 days of immersion. For XRD analysis, controlled and deteriorated mortar samples were ground, passed through a 200-μm sieve. The XRD test was conducted using the RINT D/max 2500 X-ray diffractometer at a voltage of 40 kV, current of 30 mA and a scanning speed of 2°/min using CuKα X-rays with a wavelength of 1.54 Å. For SEM and EDX analyses, samples were prepared according to Sarkar et al.'s work [27]. TGA and DSC analysis were conducted on the powdered samples obtained from sliced samples as shown in Figure 2a. The instrument used was Thermogravimetric Analyzer and Differential Scanning Calorimeter, DSC SDT Q600 (TA Instruments, New Castle, DE, USA). Samples were heated from room temperature to 1000 °C at the rate of 10 °C/min, nitrogen was purged at a flow rate of 100 mL/min. Powdered samples were prepared according to Lothenbach et al.'s work [28].

3. Results and Discussion

3.1. Compressive, Flexural Strengths and Water Absorption

Figure 3 shows the variation of compressive and flexural strengths with the addition of nano-TiO_2 in OPC and slag-blended mortars. Adding more than 6% nan-TiO_2 as weight of the binder led to a reduction in compressive strength of mortars while adding more than 3% nan-TiO_2 as weight of the binder led to a reduction in flexural strength of mortars. Increase in compressive strength was due to filler effect of nano-TiO_2 [29]. The water absorption results are shown in Figure 4 which shows that water absorption is reduced with the addition of nano-TiO_2 in both type of mortars. This positive effect can be ascribed to the nano-TiO_2 particles, which acted as fillers, served as nucleation sites for the hydration reaction, hydration products precipitated around them and disconnected pores in the paste. The reduction in water absorption due to slag addition can be ascribed to the their later hydration and formation of secondary CSH in capillary pores [30].

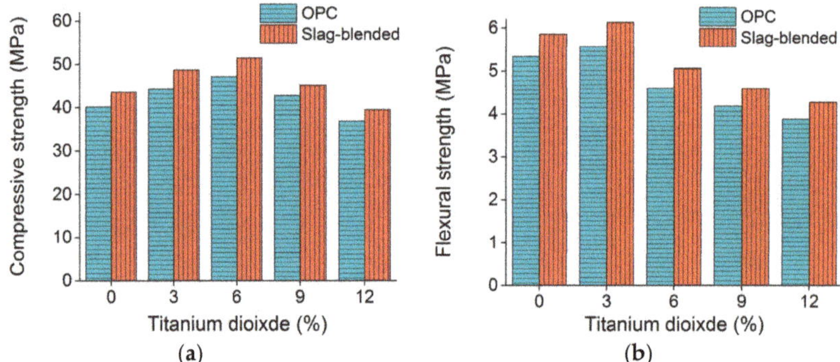

Figure 3. Variation of mechanical properties with the addition of nano-TiO_2 in mortars at the age of 28 days (**a**) compressive strength (**b**) flexural strength.

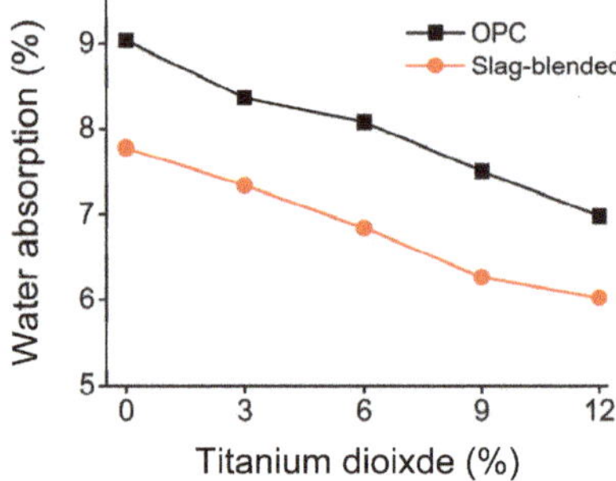

Figure 4. Variation of water absorption with the addition of nan-TiO_2 in the mortars.

3.2. Expansion and Mass Variation

Figures 5 and 6 show the expansion in nano-TiO$_2$ containing OPC and slag-blended mortars immersed in Na$_2$SO$_4$ solution at 25 °C and 5 °C for 84 days, respectively. The results show an increase in expansion rate with the addition of nano-TiO$_2$. At 25 °C, the expansion in 0% OPC mortars was 0.30%, while the expansion in 3%, 6% and 9% OPC mortars was 0.36%, 0.43% and 0.54%, respectively, after 84 days. The increase in expansion with the increase of nano-TiO$_2$ occurred at both temperatures (25 and 5 °C). The comparison of expansion in OPC and slag-blended mortars showed that nano-TiO$_2$ containing OPC mortars expanded more than nano-TiO$_2$ slag-blended mortars. The expansion pattern differed between nano-TiO$_2$ slag-blended and nano-TiO$_2$ containing OPC mortars; whereas former exhibited a uniform rate of expansion but OPC mortars showed slow expansion initially and faster expansion at later times. Slag improves performance by acting as a physical filler and pozzolanic material [31]. Therefore nano-TiO$_2$ containing slag mortars were more resistant to expansion. Storage temperature also showed a prominent effect on the rate of expansion of TiO$_2$ containing mortars.

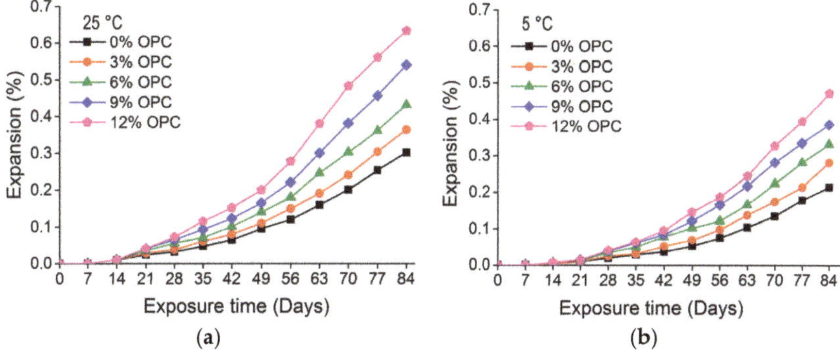

Figure 5. Expansion in nano-TiO$_2$ containing OPC mortars immersed in 10% Na$_2$SO$_4$ solution at (a) 25 °C (b) 5 °C.

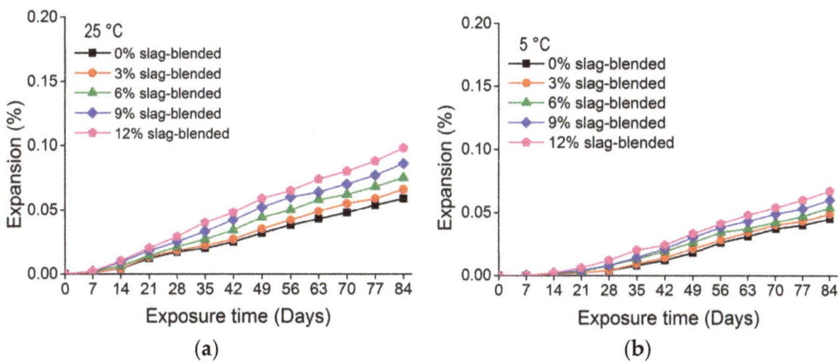

Figure 6. Expansion in nano-TiO$_2$ containing slag-blended mortars immersed in 10% Na$_2$SO$_4$ solution at (a) 25 °C (b) 5 °C.

Figures 7 and 8 show the mass variation for nano-TiO$_2$ containing OPC and slag-blended mortars immersed in Na$_2$SO$_4$ solution at 25 and 5 °C, respectively. The addition of TiO$_2$ in mortars resulted in an additional mass gain. After immersion for 84 days at 25 °C, 12% OPC and 12% slag-blended mortars gained 3.01% and 1.41% mass, respectively, while 0% OPC and 0% slag-blended mortars gained 2.19% and 1.06% mass, respectively. The increase in mass of mortars may be due to absorption of sodium sulfate solution and precipitation of gypsum and ettringite in pores and cracks. The effects of TiO$_2$ addition on the rate of mass increase were negligible in first few weeks but prominent at later stages. The increase in mass was higher in mortars immersed in sulfate solution at higher temperatures. The addition of slag refined pores and reduced capillary pores and their connectivity [6,32,33], hence reducing the absorption of sodium sulfate solution. Therefore, nano-TiO$_2$ containing slag-blended mortars were less vulnerable to the formation of gypsum and ettringite and exhibited less expansion and variation in mass.

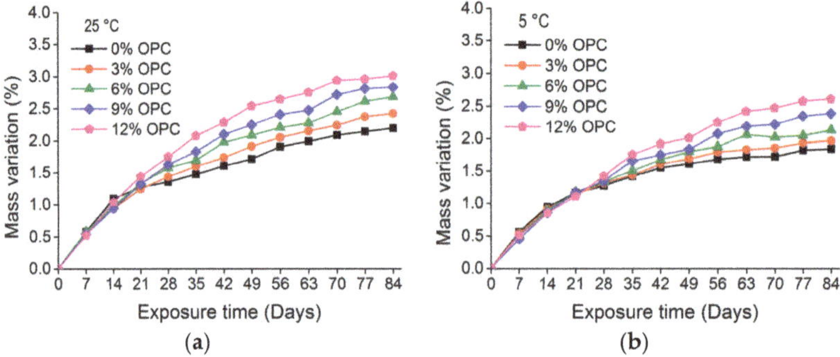

Figure 7. Mass variation in nano-TiO$_2$ containing OPC mortars immersed in 10% Na$_2$SO$_4$ solution at (**a**) 25 °; (**b**) 5 °C.

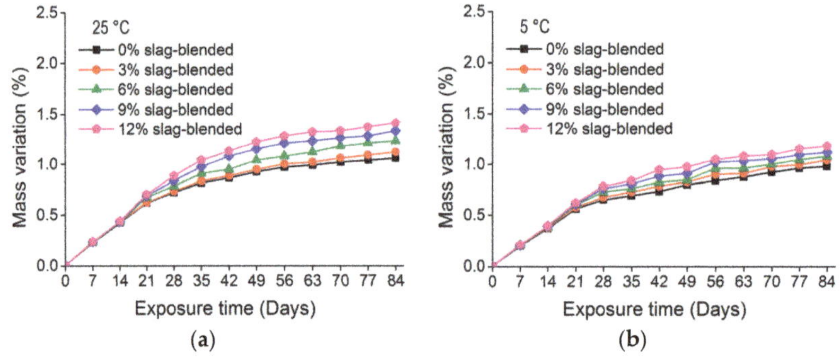

Figure 8. Mass variation in nano-TiO$_2$ containing slag-blended mortars immersed in 10% Na$_2$SO$_4$ solution at (**a**) 25 °C; (**b**) 5 °C.

3.3. Microhardness

The variation in Vickers hardness in OPC and slag-blended mortars immersed in 10% Na$_2$SO$_4$ solution at 25 °C is shown in Figure 9. It shows a greater loss of surface hardness with the addition of nano-TiO$_2$. This figure also shows that sulfate attack caused a greater loss of surface hardness in nano-TiO$_2$ containing OPC mortars than in nano-TiO$_2$ containing slag-blended mortars. In OPC mortars, most of the depth is affected and the peripheral area was softened with respect to the inner depth, while in slag-blended mortars, the loss of surface hardness value extended up to 3 mm depth. This indicates the resistance of nano-TiO$_2$ containing slag-blended mortars to sulfate attack and an increase in sulfate damage due to the addition of nanoparticles.

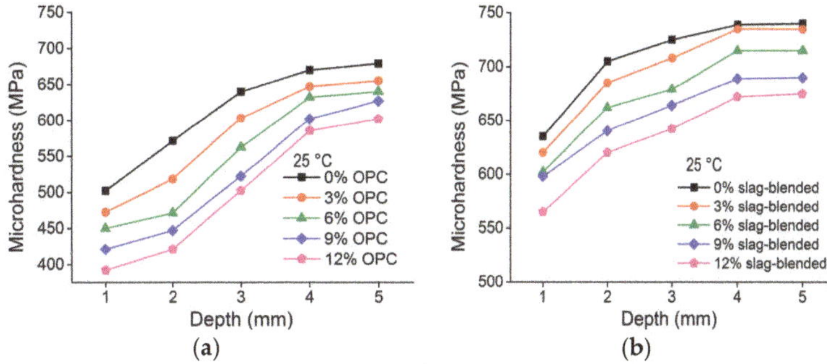

Figure 9. Variation in microhardness along depth in (**a**) nano-TiO$_2$ containing OPC mortars and (**b**) nano-TiO$_2$ containing slag-blended mortars immersed in 10% Na$_2$SO$_4$ solution at 25 °C.

3.4. Microcracks

Various cracks were formed on the surface of the specimens with respect to the axis of specimens, including perpendicular, diagonal and parallel. These cracks were referred to as transverse, diagonal and longitudinal, respectively. Cracks that changed direction from transverse to diagonal or longitudinal were identified as random cracks. Map cracks, which run in different directions and form hexagonal and irregular patterns on the surface of mortar, were also formed. At some points, spalling had occurred in the form of pop-outs and conical depressions due to the surrounding of an aggregate, or layer of the mortar, by a map crack. A representative view of the surface cracks formed on the specimens is shown in Figures 10 and 11, all of which occurred as a result of sulfate attack. Due to sulfate attack, transverse and diagonal cracks were numerous and wider than other types of cracks, followed in severity by map cracks and longitudinal cracks. This suggests that expansion of mortar samples is the dominant damage caused by sulfate attack. Table 3 shows the damage rating of nano-TiO$_2$ containing mortars on a scale from 0 to 7. The details of scale can be seen in the footnote of table. Severity of cracks varied among mix proportions. For example, 12% and 9% OPC mortars were severly damaged after 84 days compared to 3% and 0% OPC mortars. The same pattern was found in slag-blended mortars.

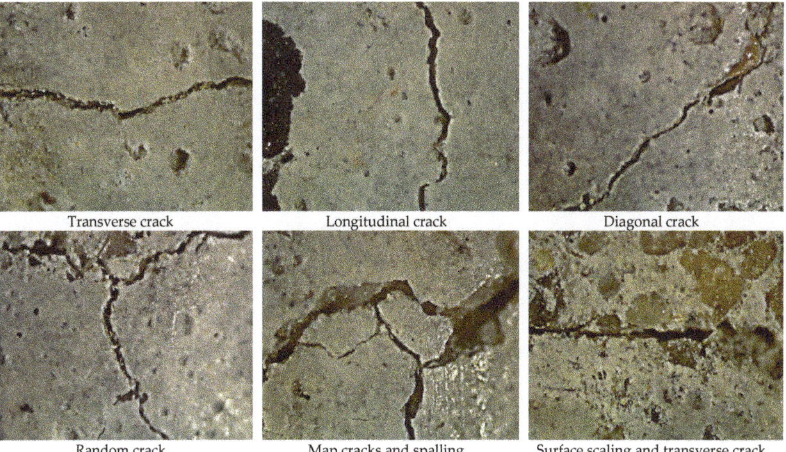

Figure 10. A representative view of cracks formed on the surface of 12% OPC mortars observed under a light microscope after 84 days of immersion in Na$_2$SO$_4$ solution at 25 °C.

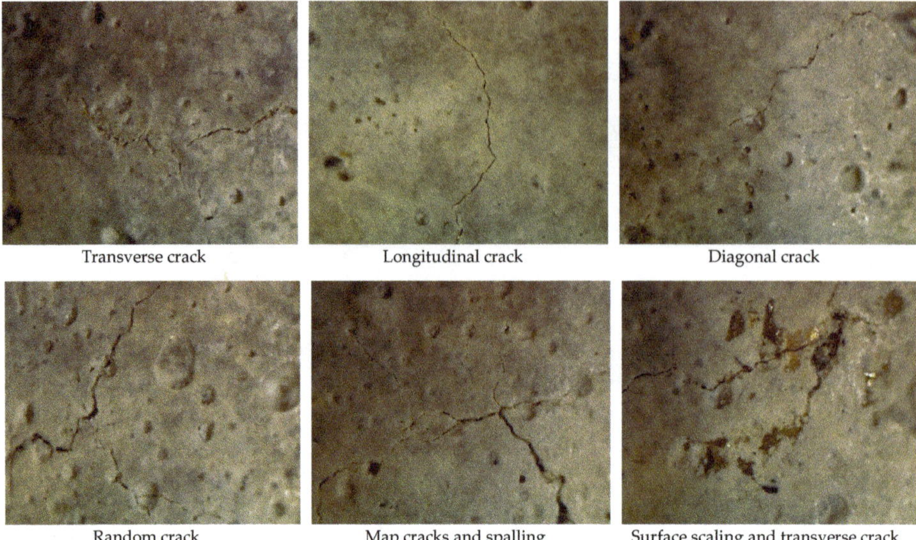

Figure 11. A representative view of cracks formed on the surface of 12% slag-blended mortars observed under a light microscope after 84 days of immersion in Na_2SO_4 solution at 25 °C.

Table 3. Damage rating in nano-TiO_2 containing OPC and slag-blended mortar specimens stored at 25 and 5 °C in Na_2SO_4 solution.

Name	25 °C				5 °C			
	0 Day	28 Days	56 Days	84 Days	0 Day	28 Days	56 Days	84 Days
0% OPC	0	1	3	4	0	1	2	3
3% OPC	0	1	3	4	0	1	2	4
6% OPC	0	1	3	4	0	1	3	4
9% OPC	0	2	3	5	0	1	3	4
12% OPC	0	2	4	6	0	2	3	5
0% slag-blended	0	0	1	2	0	0	1	2
3% slag-blended	0	0	1	2	0	0	1	2
6% slag-blended	0	0	1	2	0	0	1	2
9% slag-blended	0	0	2	3	0	0	1	2
12% slag-blended	0	0	2	3	0	0	1	2

0: No crack visible; 1: Slight cracks; 2: Some cracks; 3: Moderate cracks; 4: Severe cracks and some map cracks; 5: Severe cracks and moderate map cracks; 6: Severe cracks and minor spalling; 7: Intensive cracks and severe spalling.

3.5. XRD, SEM, EDS, TGA and DSC Analyses

Figures 12 and 13 show XRD graphs of the 0% and 12% OPC mortar and 0% and 12% slag-blended mortars stored in saturated limewater and Na_2SO_4 solution after 84 days of immersion at 25 °C, respectively. Ettringite, portlandite, quartz, calcite, anatase (TiO_2), rutile, monosulfate and hydrated tetra calcium aluminate were observed in XRD graphs of 0% and 12% OPC and slag-blended mortars before immersion, as shown in Figure 12a,c and Figure 13a,c. Ettringite is a hydration product formed due to the reaction of gypsum present in cement clinker and forms when concrete is in a plastic state. Its formation provides dimensional stability to the mortar, as it compensates for chemical shrinkage. However, its formation in later stages in hardened concrete is detrimental [34]. The peaks of portlandite in the XRD pattern of slag-blended mortars (Figure 13a,c) were lower than OPC mortars (Figure 12a,c) due to the reaction of slag with portlandite, which consumed them during a pozzolanic reaction. XRD patterns of damaged OPC (Figure 12b,d) and slag-blended mortars (Figure 13b,d) showed new peaks of ettringite and gypsum after immersion in sulfate solution, existing peaks of ettringite were increased and portlandite peaks were decreased in intensity. The appearance of new peaks of ettringite and gypsum confirmed that expansion and cracking were due to the formation of ettringite and gypsum. Although slag-blended mortars had more Al_2O_3 content than OPC mortars but they showed resistance to sulfate attack than OPC mortars. In the case of hydration of slag, Al^{3+} is bound in calcium silicate gel as C-A-S-H which shelters it form attack of incoming sulfate ions [35]. Furthermore, less monosulfate is produced in slag mortars compared to OPC mortars [13]. Monosulfate serves as an important source of Al^{3+} ions for the formation of ettringite after attack of incoming sulfate ions [13]. Therefore, less ettringite is produced in slag blended mortars.

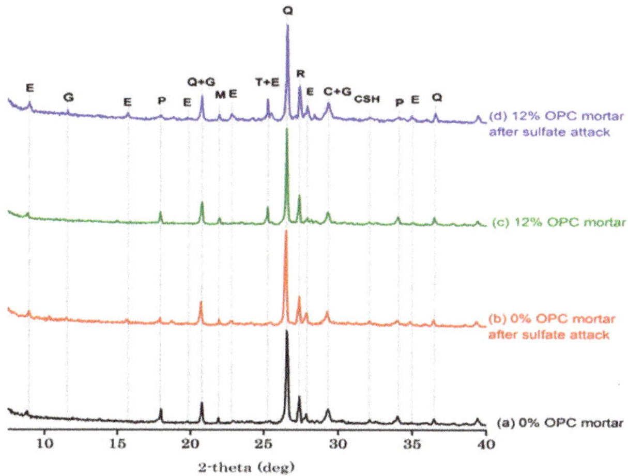

Figure 12. XRD pattern of OPC mortars (E = ettringite, G = gypsum, P = portlandite, Q = quartz, M = monosulfate, T = TiO_2 (anatase), R = rutile, C = calcite, CSH = calcium-silicate-hydrate).

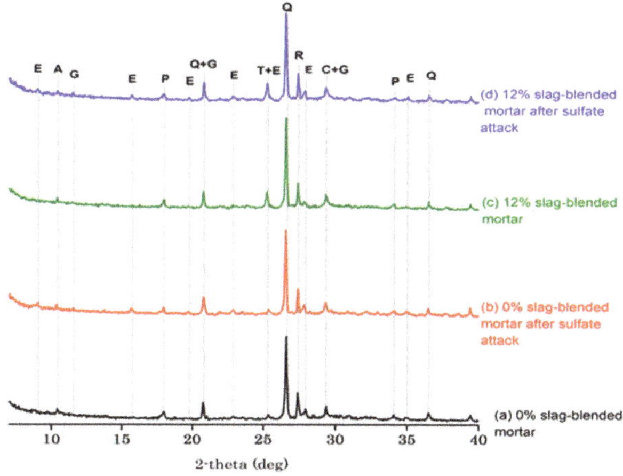

Figure 13. XRD pattern of slag-blended mortars (E = ettringite, A = tetra calcium aluminate hydrate, G = gypsum, P = portlandite, Q = quartz, T = TiO_2 (anatase), R = rutile, C = calcite, CSH = calcium-silicate-hydrate).

SEM investigations were conducted at the low and high resolution on the deteriorated specimens to investigate the internal microstructure deterioration. In the SEM image shown in Figure 14, aggregate and cement paste can be seen. The gaps around aggregates are cracks and expansion in the interfacial transition zone between the aggregate and cement paste. The cracks around the aggregates suggest that a sulfate attack caused expansion in the interfacial transition zone and damaged the bond between the aggregate and the paste. Cracks around aggregates in the outer zone are more prominent than cracks around aggregates in the inner zone. This suggests that the deterioration in samples is concentric. As samples were cylindrical, the peripheral zone was attacked first; expansive forces were then generated near the surface, causing cracks around aggregates and in the cement paste. These microcracks permitted the penetration of more sulfate ions into the microstructure. As incoming sulfate ions reacted with interior hydration products and other susceptible compounds, new crystals of ettringite and gypsum were formed in the air voids and in the cracks. As a result, the peripheral zone exhibited greater deterioration than the core. The gaps were likely filled with ettringite and gypsum. Figures 15 and 16 show needle-like crystals of ettringite formed in the pores of cement paste and plate-like crystals of gypsum. EDS analysis conducted on the ettringite and gypsum crystals is also shown in the Figures 15 and 16. A crack around the gypsum crystals is given in Figure 16. This crack is likely formed because of the crystallization pressure of gypsum.

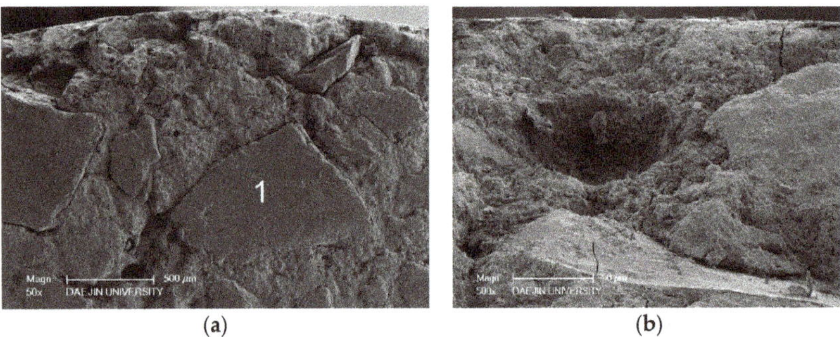

(a)   (b)

Figure 14. SEM images of 12% OPC mortar after sulfate attack (**a**) Cracks are more visible in the peripheral zone compared to the inner zone, visible crack on the outer side of aggregate 1 compared to its inner side. (**b**) A crack originating from the outer surface moving toward the inner core.

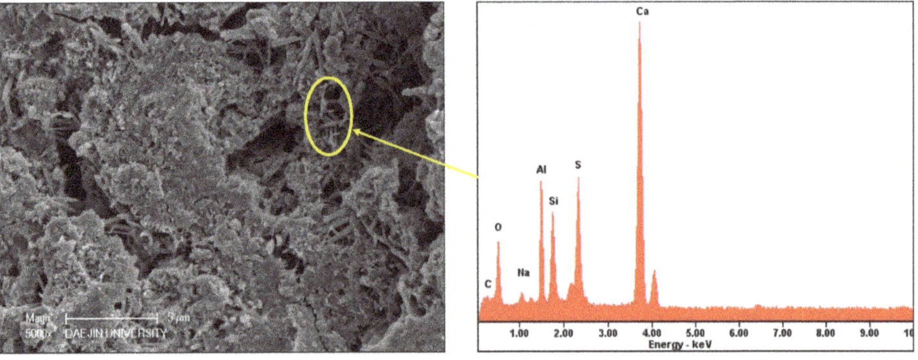

Figure 15. SEM image and EDS analyses of ettringite crystals formed in pores of 12% OPC mortar immersed at 25 °C in 10% Na_2SO_4 solution.

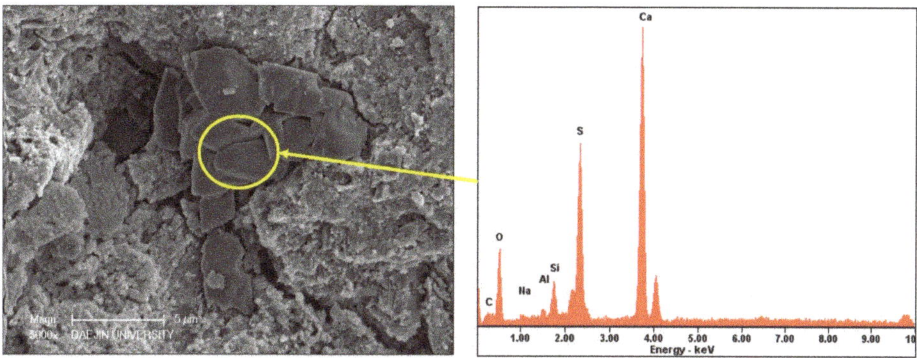

Figure 16. SEM image and EDS analyses of gypsum in 12% OPC mortar immersed at 25 °C in 10% Na$_2$SO$_4$ solution.

Thermogravimetric analysis (TGA) and differential scanning calorimetry (DSC) analysis were conducted on 0%, 6%, 12% OPC and slag blended mortars stored in saturated limewater and sodium sulfate solution at 25 °C. Differential thermogravimetry (DTG) and differential scanning calorimetry (DSC) curves for 12% OPC mortars stored in saturated limewater and Na$_2$SO$_4$ solution at 25 °C for 84 days is shown in Figure 17.

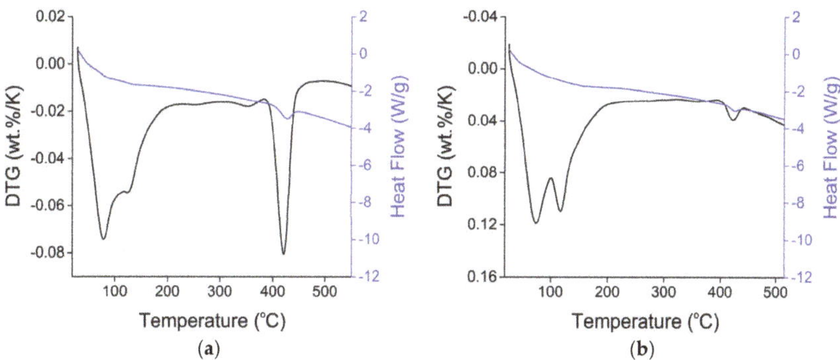

Figure 17. DTG and DSC curves in 12% OPC mortars stored in (**a**) saturated limewater; (**b**) 10% Na$_2$SO$_4$ solution.

In Figure 17a, DTG curve shows two large peak peaks at 80 °C and 430 °C, these decompositions peaks are due to the dehydration of ettringite and portlandite respectively [28]. But after exposure to sulfate attack, the peak corresponding to the portlandite has been significantly reduced, a new peak has emerged at about 130 °C. This new peak is due to the decomposition of gypsum which was produced during sulfate attack. These new peaks can also be observed in XRD shown in Figure 12d. Figure 18 shows the percentage of portlandite in mortars stored in saturated limewater (controlled mortars) and Na$_2$SO$_4$ solution (damaged mortars) [28]. This figure shows that after sulfate attack, reduction in CH content is proportional to the increasing amount of nano-TiO$_2$. Enthalpy of ettringite and gypsum corresponding to temperature 80–150 °C and portlandite corresponding to 400–500 °C was calculated from DSC curves [36] and is shown in Figure 19. A decreasing trend in enthalpy of CH in mortar samples after exposure to sulfate solution can be observed with the addition of nano-TiO$_2$. Enthalpy of ettringite and gypsum are shown in Figure 19b. Contrary to CH enthalpy, here an increase in enthalpy can be observed with the addition of nano-TiO$_2$ after sulfate attack.

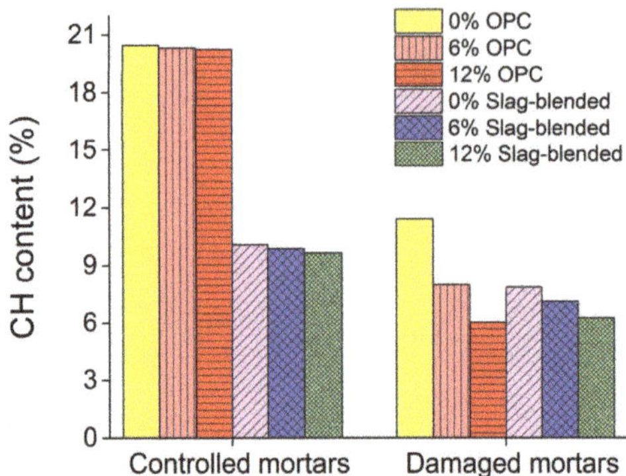

Figure 18. Portlandite content (%) in mortars stored in controlled and aggressive conditions.

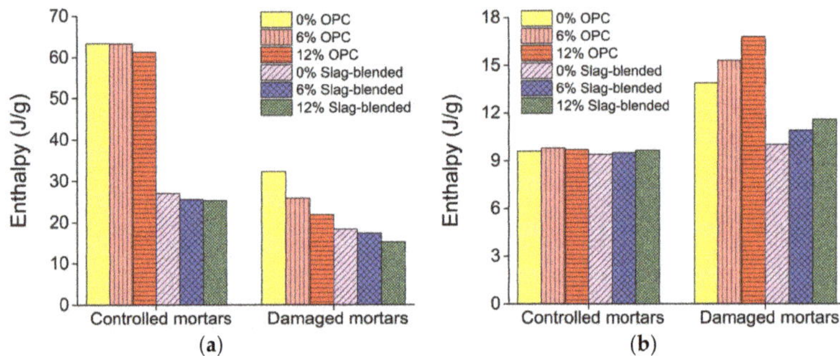

Figure 19. DSC results for (**a**) portlandite peaks at about 440 °C; (**b**) ettringite and gypsum peaks at 90–140 °C.

Crystallization pressure is considered to be responsible for the mechanism of expansion in sulfate attack [9]. When ettringite or gypsum crystals grow in a confined space, they exert an expansive force on the walls of the pores. According to Scherer [37], crystallization pressure is inversely proportional to the size of the pores. If the pores in which crystals grow are smaller in size, then the crystallization pressure will be higher. When the crystallization pressure is greater than the tensile strength of the mortar, the pores will expand and microcracks will occur. After microcracking, more sulfate ions will enter the sample, reacting with portlandite and forming gypsum.

TiO_2 is an inert material [14,38], it does not react with water, cement and incoming sulfate ions. The nano-TiO_2 particles act as nuclei for hydration reaction. The hydration products grow around them and fill the voids, thus reduce the porosity [14]. Mercury intrusion porosimetry (MIP) studies of nano-TiO_2 containing cementitious materials [14,39–42] have shown that the addition of nano-TiO_2 have refined their pores and the most probable pore diameters of cementitious materials have shifted towards smaller pore diameters. As stated above that during sulfate attack, expansive products (ettringite and gypsum) are created in the pores of the cement mortar and they create crystallization pressure on the walls of pores. If the size of pores is smaller, the crystallization pressure will be higher and vice versa. Here, the pore sizes were reduced due to the addition of nano-TiO_2, so the crystallization pressure was higher due to the formation of ettringite and gypsum crystals in pores of nano-TiO_2 containing mortars than control mortars. Secondly, the addition of nano-TiO_2 particles can reduce the tensile strength of mortar [15]; therefore, the addition of nano-TiO_2 not only increased crystallization pressure, which is a tensile force but also simultaneously reduced tensile strength, leaving mortars with nano-TiO_2 more vulnerable to sulfate attack and salt crystallization pressure than normal mortars. According to Santhanam et al. [43], the presence of voids can reduce expansive stresses due to crystallization, which can reduce the number of cracks in the paste. The researchers [43] observed a slower rate of disintegration in the air-entrained mortars than in the non-air-entrained mortars. Therefore, nano-TiO_2 containing mortars had less voids and more expansive stresses created due to crystallization pressure in smaller size voids and were ultimately more damaged than mortars without nano-TiO_2. Lee and Kurtis [44] also concluded that the addition of nanoparticles increases the salt crystallization pressure. In their study, mortars containing nanoparticles showed greater damage due to salt crystallization than mortars without nanoparticles. TiO_2 containing slag-blended mortars were resistant to sulfate attack because slag reacts with portlandite to form calcium silicate hydrate [31]. It consumes portlandite and increases the amount of CSH in mortars. Additional hydrates are formed in larger capillary pores and reduce pore connectivity [45]. This process likely reduced the porosity and permeability of TiO_2 containing slag-blended mortars and increased the compressive and tensile strength of mortars [6]. The lower content of portlandite in slag-blended mortars can be observed in the XRD pattern shown in Figure 13. The other reactive hydration products susceptible to sulfate attack are monosulfoaluminate hydrate, calcium aluminate hydrate and calcium aluminate hydroxide. These products were likely in lower quantities in slag mortars due to the absence of C_3A clinker phases in slag and these products are less reactive when present in slag [36]. As these phases are required for the formation of ettringite, their lower quantities in slag are not sufficient to cause ettringite-related expansion and cracking. The higher tensile strength of slag mortars (compared to OPC mortars) can also resist the expansion due to the formation of ettringite so nano-TiO_2 containing slag-blended mortars were more resistant to sulfate attack than so nano-TiO_2 containing OPC mortars.

4. Conclusions

In this study, we investigated the influence of titanium dioxide (TiO_2) nanoparticles on the sulfate attack upon ordinary Portland cement and slag-blended mortars. Nano-TiO_2 containing mortars were immersed in 10% Na_2SO_4 solution at 25 and 5 °C. The expansion, variation in mass and surface microhardness were observed. Deteriorated samples were analyzed using XRD, SEM, EDS, TGA and DSC tests. Based on the test results, the following points were concluded.

Deterioration due to sulfate attack occurred on all TiO_2 containing mortars. The results show that the addition of TiO_2 has robust effects on the rate of sulfate attack on OPC and slag-blended mortars. The extent of damage increases with the addition of TiO_2 nanoparticles. It is recommended that mortars and concrete structures containing TiO_2 for self-cleaning and air purifying purposes should be designed as sulfate resistant.

The TiO_2 containing slag-blended mortars showed lower expansion, mass and surface hardness changes and cracking. So, blast furnace slag can be used to reduce sulfate attack in nano-TiO_2 containing mortars.

The TiO_2 containing mortar and concrete structures exposed to higher temperatures are more vulnerable to sulfate attack as the rate of expansion, mass variation and cracks were higher in mortars stored at 25 °C temperature than at 5 °C. Specimens at lower temperatures take more time to match the expansion of specimens at higher temperatures.

The rate and severity of sulfate attack on TiO_2 containing cementitious materials should be investigated by varying the factors like the type of cations of sulfate solution (K^+, Mg^{2+}, Ca^{2+}), pH of the solution and size of the specimens. Also, the influence of titanium dioxide on sulfate attack upon other materials such as geopolymer concrete should be investigated.

Acknowledgments: This study was supported by the Technology Development Program (C0532231) funded by the Ministry of SMEs and Startups (MMS, Korea).

Author Contributions: Atta-ur-Rehman, Abdul Qudoos, and Hong Gi Kim conceived, designed, performed the experiments and analyzed the data under the guidance and supervision of Jae-Suk Ryou.

Conflicts of Interest: The authors have no conflict of interest.

References

1. Chen, J.; Poon, C.-S. Photocatalytic construction and building materials: From fundamentals to applications. *Build. Environ.* **2009**, *44*, 1899–1906. [CrossRef]
2. Chen, J.; Kou, S.-C.; Poon, C.-S. Photocatalytic cement-based materials: Comparison of nitrogen oxides and toluene removal potentials and evaluation of self-cleaning performance. *Build. Environ.* **2011**, *46*, 1827–1833. [CrossRef]
3. Strini, A.; Roviello, G.; Ricciotti, L.; Ferone, C.; Messina, F.; Schiavi, L.; Corsaro, D.; Cioffi, R. TiO$_2$-based photocatalytic geopolymers for nitric oxide degradation. *Materials* **2016**, *9*, 513. [CrossRef] [PubMed]
4. Ohama, Y.; Van Gemert, D. *Application of Titanium Dioxide Photocatalysis to Construction Materials: State-of-the-Art Report of the Rilem Technical Committee 194-TDP*; Springer Science & Business Media: New York, NY, USA, 2011; Volume 5.
5. Hüsken, G.; Hunger, M.; Brouwers, H. Experimental study of photocatalytic concrete products for air purification. *Build. Environ.* **2009**, *44*, 2463–2474. [CrossRef]
6. Siddique, R.; Khan, M.I. *Supplementary Cementing Materials*; Springer Science & Business Media: New York, NY, USA, 2011.
7. Virgalitte, S.J.; Luther, M.D.; Rose, J.H.; Mather, B.; Bell, L.W.; Ehmke, B.A.; Klieger, P.; Roy, D.M.; Call, B.M.; Hooton, R.D. *Ground Granulated Blast-Furnace Slag as a Cementitious Constituent in Concrete*; ACI Report 233R-95; American Concrete Institute: Detroit, MI, USA, 1995.
8. Bastos, G.; Patiño-Barbeito, F.; Patiño-Cambeiro, F.; Armesto, J. Nano-inclusions applied in cement-matrix composites: A review. *Materials* **2016**, *9*, 1015. [CrossRef] [PubMed]
9. Yu, C.; Sun, W.; Scrivener, K. Mechanism of expansion of mortars immersed in sodium sulfate solutions. *Cem. Concr. Res.* **2013**, *43*, 105–111. [CrossRef]
10. Gruyaert, E.; Van den Heede, P.; Maes, M.; De Belie, N. Investigation of the influence of blast-furnace slag on the resistance of concrete against organic acid or sulphate attack by means of accelerated degradation tests. *Cem. Concr. Res.* **2012**, *42*, 173–185. [CrossRef]
11. Santhanam, M.; Cohen, M.D.; Olek, J. Sulfate attack research—Whither now? *Cem. Concr. Res.* **2001**, *31*, 845–851. [CrossRef]
12. Neville, A. The confused world of sulfate attack on concrete. *Cem. Concr. Res.* **2004**, *34*, 1275–1296. [CrossRef]
13. Marchand, J.; Odler, I.; Skalny, J.P. *Sulfate Attack on Concrete*; CRC Press: New York, NY, USA, 2003.
14. Chen, J.; Kou, S.-C.; Poon, C.-S. Hydration and properties of nano-TiO$_2$ blended cement composites. *Cem. Concr. Compos.* **2012**, *34*, 642–649. [CrossRef]
15. Lucas, S.; Ferreira, V.; de Aguiar, J.B. Incorporation of titanium dioxide nanoparticles in mortars—Influence of microstructure in the hardened state properties and photocatalytic activity. *Cem. Concr. Res.* **2013**, *43*, 112–120. [CrossRef]
16. Meng, T.; Yu, Y.; Qian, X.; Zhan, S.; Qian, K. Effect of nano-TiO$_2$ on the mechanical properties of cement mortar. *Constr. Buil. Mater.* **2012**, *29*, 241–245. [CrossRef]
17. Lee, B.Y. *Effect of Titanium Dioxide Nanoparticles on Early Age and Long Term Properties of Cementitious Materials*; Georgia Institute of Technology: Atlanta, Georgia, 2012.
18. Pérez-Nicolás, M.; Balbuena, J.; Cruz-Yusta, M.; Sánchez, L.; Navarro-Blasco, I.; Fernández, J.M.; Alvarez, J.I. Photocatalytic NO$_X$ abatement by calcium aluminate cements modified with TiO$_2$: Improved NO$_2$ conversion. *Cem. Concr. Res.* **2015**, *70*, 67–76. [CrossRef]
19. *Standard Test Method for Length Change of Hydraulic-Cement Mortars Exposed to a Sulfate Solution American*; ASTM C1012/C1012M-18a; American Society for Testing and Materials: West Conshohocken, PA, USA, 2018.
20. Van Tittelboom, K.; De Belie, N.; Hooton, R.D. Test methods for resistance of concrete to sulfate attack—A critical review. In *Performance of Cement-Based Materials in Aggressive Aqueous Environments*; Springer: New York, NY, USA, 2013; pp. 251–288.
21. Ferraris, C.; Stutzman, P.; Peltz, M.; Winpigler, J. Developing a more rapid test to assess sulfate resistance of hydraulic cements. *J. Res. Natl. Inst. Stand. Technol.* **2005**, *110*, 529. [CrossRef] [PubMed]
22. Ferraris, C.F.; Clifton, J.R.; Stutzman, P.E.; Garboczi, E. *Mechanisms of Chemical Degradation of Cement-Based Systems*; CRC Press: New York, NY, USA, 1997; p. 185.

23. *Standard Test Method for Compressive Strength of Hydraulic Cement Mortars (Using 2-in. or [50-mm] Cube Specimens)*; ASTM C109/C109M-16a; American Society for Testing and Materials: West Conshohocken, PA, USA, 2016.

24. *Standard Test Method for Flexural Strength and Modulus of Hydraulic Cement Mortars*; ASTM, C348-02; American Society for Testing and Materials: West Conshohocken, PA, USA, 2002.

25. *Standard Test Method for Density, Absorption and Voids in Hardened Concrete*; ASTM, C642-13; American Society for Testing and Materials: West Conshohocken, PA, USA, 2013.

26. Igarashi, S.; Bentur, A.; Mindess, S. Microhardness testing of cementitious materials. *Adv. Cem. Based Mater.* **1996**, *4*, 48–57. [CrossRef]

27. Sarkar, S.L.; Aimin, X.; Jana, D. Scanning electron microscopy, X-ray microanalysis of concretes-7. In *Handbook of Adhesives Raw Materials*; Noyes Publications: Park Ridge, NJ, USA, 2001.

28. Lothenbach, B.; Durdzinski, P.; De Weerdt, K. Thermogravimetric analysis. In *A Practical Guide to Microstructural Analysis of Cementitious Materials*; CRC Press: Oxford, UK, 2016; pp. 177–212.

29. Nazari, A.; Riahi, S. The effect of TiO_2 nanoparticles on water permeability and thermal and mechanical properties of high strength self-compacting concrete. *Mater. Sci. Eng. A* **2010**, *528*, 756–763. [CrossRef]

30. Hooton, R.D. Canadian use of ground granulated blast-furnace slag as a supplementary cementing material for enhanced performance of concrete. *Can. J. Civ. Eng.* **2000**, *27*, 754–760. [CrossRef]

31. Mehta, P.K. *Concrete: Structure, Properties and Materials*; Prentice Hall: Upper Saddle River, NJ, USA, 1986.

32. Bouikni, A.; Swamy, R.N.; Bali, A. Durability properties of concrete containing 50% and 65% slag. *Constr. Build. Mater.* **2009**, *23*, 2836–2845. [CrossRef]

33. Lang, E. Blast furnace cements. In *Structure and Performance of Cements*, 2nd ed.; Bensted, J., Barnes, P., Eds.; Spon Press: London, UK, 2002; pp. 310–325.

34. Menéndez, E.; Matschei, T.; Glasser, F.P. Sulfate attack of concrete. In *Performance of Cement-Based Materials in Aggressive Aqueous Environments*; Springer: New York, NY, USA, 2013; pp. 7–74.

35. Whittaker, M.; Zajac, M.; Ben Haha, M.; Black, L. The impact of alumina availability on sulfate resistance of slag composite cements. *Constr. Build. Mater.* **2016**, *119*, 356–369. [CrossRef]

36. Bassuoni, M.T.; Nehdi, M.L. Durability of self-consolidating concrete to sulfate attack under combined cyclic environments and flexural loading. *Cem. Concr. Res.* **2009**, *39*, 206–226. [CrossRef]

37. Scherer, G.W. Crystallization in pores. *Cem. Concr. Res.* **1999**, *29*, 1347–1358. [CrossRef]

38. Jayapalan, A.; Lee, B.; Fredrich, S.; Kurtis, K. Influence of additions of anatase TiO_2 nanoparticles on early-age properties of cement-based materials. *Transp. Res. Rec. J. Transp. Res. Board* **2010**, *2141*, 41–46. [CrossRef]

39. Yang, L.Y.; Jia, Z.J.; Zhang, Y.M.; Dai, J.G. Effects of nano-TiO_2 on strength, shrinkage and microstructure of alkali activated slag pastes. *Cem. Concr. Compos.* **2015**, *57*, 1–7. [CrossRef]

40. Zhang, R.; Cheng, X.; Hou, P.; Ye, Z. Influences of nano-TiO_2 on the properties of cement-based materials: Hydration and drying shrinkage. *Constr. Build. Mater.* **2015**, *81*, 35–41. [CrossRef]

41. Jimenez-Relinque, E.; Rodriguez-Garcia, J.R.; Castillo, A.; Castellote, M. Characteristics and efficiency of photocatalytic cementitious materials: Type of binder, roughness and microstructure. *Cem. Concr. Res.* **2015**, *71*, 124–131. [CrossRef]

42. Nazari, A.; Riahi, S. TiO_2 nanoparticles effects on physical, thermal and mechanical properties of self compacting concrete with ground granulated blast furnace slag as binder. *Energy Build.* **2011**, *43*, 995–1002. [CrossRef]

43. Santhanam, M.; Cohen, M.D.; Olek, J. Mechanism of sulfate attack: A fresh look: Part 2. Proposed mechanisms. *Cem. Concr. Res.* **2003**, *33*, 341–346. [CrossRef]

44. Lee, B.Y.; Kurtis, K.E. Effect of pore structure on salt crystallization damage of cement-based materials: Consideration of w/b and nanoparticle use. *Cem. Concr. Res.* **2017**, *98*, 61–70. [CrossRef]

45. Divsholi, B.S.; Lim, T.Y.D.; Teng, S. Durability properties and microstructure of ground granulated blast furnace slag cement concrete. *Int. Concr. Struct. Mater.* **2014**, *8*, 157–164. [CrossRef]

Article

Shell Layer Thickness-Dependent Photocatalytic Activity of Sputtering Synthesized Hexagonally Structured ZnO-ZnS Composite Nanorods

Yuan-Chang Liang *, Ya-Ru Lo, Chein-Chung Wang and Nian-Cih Xu

Institute of Materials Engineering, National Taiwan Ocean University, Keelung 20224, Taiwan;
yalulo0807@gmail.com (Y.-R.L.); abc2589tw@gmail.com (C.-C.W.); sad821008@gmail.com (N.-C.X.)
* Correspondence: yuanvictory@gmail.com

Received: 13 December 2017; Accepted: 5 January 2018; Published: 7 January 2018

Abstract: ZnO-ZnS core-shell nanorods are synthesized by combining the hydrothermal method and vacuum sputtering. The core-shell nanorods with variable ZnS shell thickness (7–46 nm) are synthesized by varying ZnS sputtering duration. Structural analyses demonstrated that the as-grown ZnS shell layers are well crystallized with preferring growth direction of ZnS (002). The sputtering-assisted synthesized ZnO-ZnS core-shell nanorods are in a wurtzite structure. Moreover, photoluminance spectral analysis indicated that the introduction of a ZnS shell layer improved the photoexcited electron and hole separation efficiency of the ZnO nanorods. A strong correlation between effective charge separation and the shell thickness aids the photocatalytic behavior of the nanorods and improves their photoresponsive nature. The results of comparative degradation efficiency toward methylene blue showed that the ZnO-ZnS nanorods with the shell thickness of approximately 17 nm have the highest photocatalytic performance than the ZnO-ZnS nanorods with other shell layer thicknesses. The highly reusable catalytic efficiency and superior photocatalytic performance of the ZnO-ZnS nanorods with 17 nm-thick ZnS shell layer supports their potential for environmental applications.

Keywords: sputtering; composite nanorods; shell thickness; photocatalytic activity

1. Introduction

The high electron mobility, wide and direct band gap (3.1 eV–3.4 eV) and large exciton binding energy (60 meV) at room temperature spread various potential applications of ZnO [1]. Moreover, ZnO is also a cost-efficient and environment-friendly material; therefore, it is promising for applications to alleviate environmental problems. Photoexcited electron–hole pairs in ZnO under light irradiation can interact with the O_2 adsorbed on the surface of the ZnO photocatalyst and H_2O to generate $O_2{}^-$ and $\cdot OH$, respectively, which can reduce and oxidize the organic contaminants; the ZnO is promising for photocatalytic applications [2]. Changing the morphology of ZnO is an efficient way to enhance its photocatalytic efficiency. Various works have reported low-dimensional ZnO with different morphologies in photocatalyst applications [3,4]. In these studies, ZnO nanorods are posited to be a suitable architecture for photocatalytic applications [5]; moreover, ZnO nanorods coupled with other semiconductors to form core–shell nanostructures are a viable strategy to realize the efficient separation of photoinduced charge carriers in order to improve the photocatalytic performance of the ZnO nanorods. The TiO_2-coated ZnO nanorods, CdS-coated ZnO nanorods, and Sn_2S_3-coated ZnO nanorods are reported to show superior photocatalytic performance than that of pure ZnO nanorods [6–8]. In addition, ZnS is a wide bandgap semiconductor (3.6 eV) [9]. ZnS can be designed in many forms, like particles, thin films, wires, rods, tubes, and sheets [10]. Recent work shows that ZnS is also a promising photocatalyst [11]. It has been shown that the formation of hollow ZnO core-ZnS shell structure improves the photocatalytic activity of pristine ZnO [12]. Moreover, when compared with ZnO spheres, the photocatalytic activities for methyl orange of the ZnO-ZnS core-shell sphere structure is improved [13]. Theoretical calculations and experimental results have demonstrated that the combination of the ZnO and ZnS wide bandgap semiconductors can yield a novel material that has photo-induced threshold energy lower than that of the individual components [14]. Based on the aforementioned discussions, the integration of ZnS into ZnO nanorods is a promising approach to enhance the photocatalytic properties of the ZnO nanorods.

Covering ZnS shell layer crystallites onto the ZnO nanorods, a proper thin-film process of ZnS is crucial to fabricate ZnO-ZnS core-shell composite nanorods with a successive architecture. Several techniques, including sputtering [15], thermal evaporation [16,17], pulsed laser deposition, [18], and chemical bath deposition [19] have been used to grow ZnS thin films. In particular, the sputtering technique is the most promising method for up-scaling the growth system while maintaining a good control of the deposition rate, so it is widely used in industrial applications [20,21]. In the present study, ZnS crystallites with various thicknesses were sputtered onto ZnO nanorods to form ZnO-ZnS core-shell heterostructured nanorods. The correlation between microstructures and photocatalytic properties of the sputtering formation of ZnO-ZnS nanorods was investigated.

2. Materials and Methods

In this study, ZnO-ZnS core-shell nanorods with different ZnS shell layer thicknesses were fabricated by sputtering ZnS thin films with different sputtering durations onto the surfaces of hydrothermally derived ZnO nanorod templates. The synthesis of the ZnO nanorods consisted of two steps corresponding to the formation of a ZnO seed layer and the growth of nanorods. In the first, the ZnO seed layer was deposited on the 300 nm-thick SiO_2/Si substrate by magnetron sputtering. Subsequently, the substrates were perpendicularly suspended in a solution containing equimolar (0.05 M) aqueous solutions of zinc nitratehexahydrate ($Zn(NO_3)_2 \cdot 6H_2O$) and hexamethylenetetramine ($C_6H_{12}N_4$). The hydrothermal reaction temperature was fixed at 95 °C and the duration for crystal growth is 9 h. The as-synthesized ZnO nanorods have the average diameter of 135 nm. Moreover, the average length of the ZnO nanorods is 1.36 μm. The ZnS shell layer was deposited on the ZnO nanorods by radio-frequency (RF) magnetron sputtering of a ZnS target. ZnS film was deposited in pure Ar ambient at 80 W under 2.67 Pa working pressure. The ZnS film growth temperature was fixed at 460 °C. The thickness of the ZnS shell was varied by controlling the sputtering duration from 10 to 60 min.

Sample crystal structures of the ZnO-ZnS composite nanorods were investigated by X-ray diffraction (XRD, Bruker D2 PHASER, Karlsruhe, Germany) using Cu Ka radiation. The surface morphology of the composite nanorods was characterized by scanning electron microscopy (SEM, Hitachi S-4800, Tokyo, Japan). The crystallinity and thickness of the ZnS shell layer were measured by transmission electron microscopy (TEM, Philips Tecnai F20 G2, Amsterdam, The Netherland). The rod-like thin film samples with a thickness of approximately 1.36 μm are used for the room-temperature photoluminescence (PL) measurements. The Horiba Jobin Yvon HR 800 (Horiba Scientific, Kyoto, Japan) with the 325 nm continuous wave He-Cd laser source are used in the emission measurements. To measure photocurrent properties of the composite nanorods, solarlight irradiation excited from a 100 W Xe arc lamp was used for illumination. Silver glues were laid on the surfaces of the samples to form two contact electrodes, and the applied voltage was fixed at 5 V during electric measurements. Photocatalytic activity of various ZnO-ZnS composite nanorods were performed by comparing the degradation of aqueous solution of methylene blue (MB, 5×10^{-6} M) containing various ZnO-ZnS nanorods as catalysts under solarlight irradiation excited from a 100 W Xe arc lamp. The solution volume of MB is 10 mL and the ZnO-ZnS nanorods are grown on the substrates with a fixed coverage area of 1.2 cm × 1.2 cm for the photodegradation tests. The variation of MB solution concentration in the presence of various ZnO-ZnS nanorods with different irradiation durations was analyzed by recording the absorbance spectra using an UV-Vis spectrophotometer (Jasco V750, Tokyo, Japan).

3. Results and Discussion

Figure 1a–d show the SEM images of the as-synthesized ZnO-ZnS core–shell nanorods with various ZnS sputtering durations. Upon increasing the sputtering duration, a marked increase in the diameter of the composite nanorods was visible. In the heterostructure systems of sputtering deposited perovskite $La_{0.7}Sr_{0.3}MnO_3$ epilayers on $SrTiO_3$ substrates and $(La,Ba)MnO_3$ films on $SrTiO_3$ substrates, a thicker film engenders roughening of surface morphology [22,23]. The lattice misfit between the as-grown film and the underlayer material causes the marked thickness-dependent roughness scaling effect. The ZnS and ZnO have the same wurtize hexagonal structure, but different lattice constants. An increase in sputtering deposited ZnS shell layer thickness might cause increased layer surface undulation. Figure 2a–d exhibit XRD patterns of the ZnO-ZnS composite nanorods with various ZnS shell layer thicknesses. The intense Bragg reflection that was centered at approximately 34.4° is ascribed to the ZnO (002). No other Bragg reflection from the ZnO was distinguished revealed that the highly c-axis oriented feature of the ZnO nanorods. In addition to ZnO (002), an obvious Bragg reflection centered at approximately at 28.6° is noticed which corresponds to (002) plane of wurtzite ZnS (JCPDS No. 005-0492). The relative intensity of the ZnS (002) Bragg reflection increased upon increasing the ZnS sputtering duration and no other Bragg reflection originated from other crystallographic planes was observed. This indicates that hexagonal ZnO crystal provides well crystallographic accommodation for the growth of hexagonal ZnS shell layers by sputtering deposition at 460 °C in this study [24]. By contrast, the zinc blend cubic ZnS phase is favorable to form at a low-temperature sputtering process and the substrate type affects the crystallographic structure of ZnS [25]. It has been shown that the ZnO-ZnS composites with various morphologies composed of hexagonal ZnO and cubic ZnS are formed at other low temperature synthesis methods [12]. The sputtering growth of the ZnS shell layer herein provides ZnS adatoms with sufficient energy to adsorb onto the ZnO facets and growth with the similar crystallographic feature of the under-layered ZnO crystal at the given growth temperature [26]; therefore, the ZnO-ZnS composite nanorods with single hexagonal phase were formed.

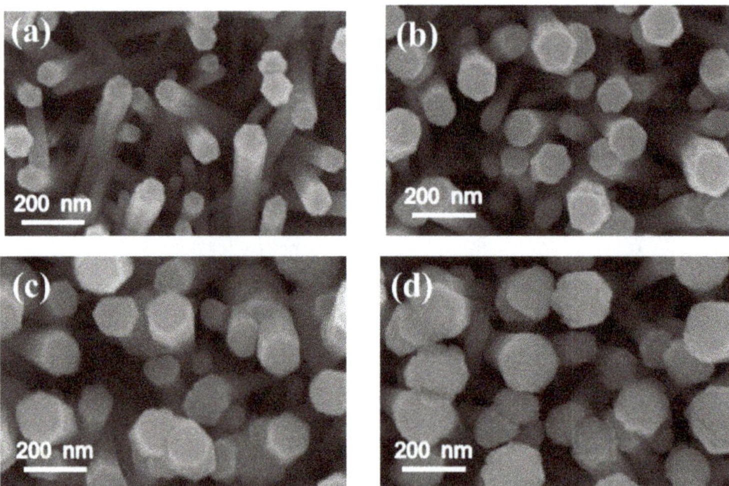

Figure 1. SEM images of ZnO-ZnS core-shell nanorods with various ZnS sputtering durations: (**a**) 10 min; (**b**) 20 min; (**c**) 40 min and (**d**) 60 min.

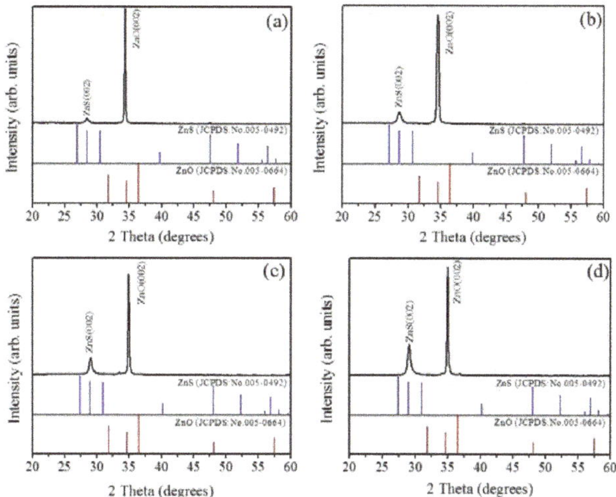

Figure 2. X-ray diffraction (XRD) patterns of ZnO-ZnS core-shell nanorods with various ZnS sputtering durations: (**a**) 10 min; (**b**) 20 min; (**c**) 40 min and (**d**) 60 min.

Figure 3a shows a low-magnification TEM image of a single ZnO-ZnS core–shell nanorod with the ZnS sputtering duration of 10 min. An ultrathin ZnS layer was homogeneously covered over the surface of the ZnO nanorod. The thickness of the ZnS shell layer is estimated to be approximately 7 nm. Figure 3b,c show the high-resolution TEM (HRTEM) images of the ZnO-ZnS core–shell nanorod taken from the different local regions at the ZnO-ZnS interface. The arrangement of lattice fringes of the shell layer is visible and highly ordered and correlated with those of the ZnO core, revealing that the atomic arrangement orientation of the ZnO and ZnS crystals is in a similar manner. The clear lattice fringes in the HRTEM images with an inter-planar spacing of approximately 0.31 nm corresponded to the (002) plane of the hexagonal ZnS structure. The arrangement of lattice fringes in Figure 3b,c demonstrates the crystal orientations of the ZnO and ZnS match well along (002) plane. Figure 3d exhibits selected area electron diffraction (SAED) pattern of the ZnO-ZnS nanorod. The SAED pattern shows that both ZnO core and ZnS shell crystals are grown along c-axis direction. Because of the similar crystal structure and the crystallographic growth orientation between the constituent compounds, the satellite spots are visible. This is associated from the reconstruction of the diffraction from the overlapping planes of the ZnO and ZnS [27]. The TEM analyses revealed that the ZnO-ZnS nanorod synthesized by sputtering ZnS shell layer has a high crystallinity. Figure 3e displays low-magnification TEM image of the ZnO-ZnS nanorod with 20 min sputtering growth of ZnS shell layer; the as-deposited ZnS shell thickness is estimated approximately 17 nm. Moreover, Figure 3f demonstrates that the ZnS shell layer still exhibited well-ordered arranged lattice fringes, revealing the highly crystallinity of the ZnS crystallites. The energy dispersive spectroscopy (EDS) spectra in Figure 3g shows that Zn, O, and S are the primary detected elements and no other impurity atoms are detected. Figure 3h,i display the low-magnification TEM images of the ZnO-ZnS nanorods with 40 min and 60 min ZnS sputtering durations; the outer gray-layer is originated from the ZnS. The thicknesses of the ZnS shell layer were estimated to around 32 nm and 46 nm, respectively. The EDS spectra of the ZnO-ZnS nanorod with a 46 nm-thick ZnS shell layer were also displayed in Figure 3j. Notably, since the ZnS shell layer thickness was thickened by a prolonged sputtering process, the intensity of sulfur signal was markedly increased compared with that shown in Figure 3g wherein the core-shell composite nanorod has a thinner ZnS shell layer.

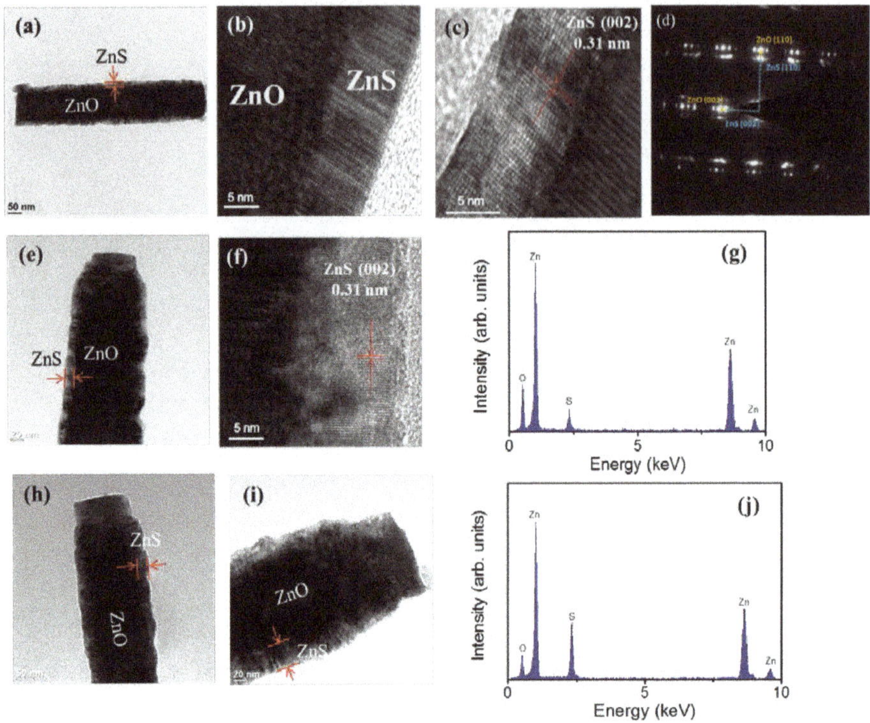

Figure 3. (**a**) Low-magnification transmission electron microscopy (TEM) imageof ZnO-ZnS composite nanorod with ZnS sputtering 10 min; the thickness of ZnS is approximately 7 nm; (**b,c**) The corresponding high-resolution images obtained from the various regions of the composite nanorod; (**d**) Selected area electron diffraction (SAED) pattern of the nanorod; (**e**) Low-magnification TEM imageof ZnO-ZnS composite nanorod with ZnS sputtering 20 min, the thickness of ZnS is approximately 17 nm; (**f**) The corresponding high-resolution image obtained from the outer region of the nanorod; (**g**) Energy dispersive spectroscopy (EDS) spectra of the nanorod; (**h,i**) Low-magnification TEM images of ZnO-ZnS composite nanorods with ZnS sputtering 40 min and 60 min; the thicknesses of ZnS are approximately 32 nm and 46 nm, respectively; (**j**) EDS spectra of the ZnO-ZnS nanorod in (**i**).

Figure 4 illustrates the PL spectra of the ZnO-ZnS core–shell nanorods with different ZnS shell layer thicknesses. A sharp and distinct UV emission band was ascribed to the near-band edge (NBE) emission of the ZnO nanorods [8]. Moreover, a broad and clear visible-light emission band centered at approximately 570 nm was observed for the ZnO nanorods. This green-yellow band for visible luminescence is due to the discrete deep energy levels in the band gap formed by the point defects [28]. Notably, the NBE emission intensity of the ZnO nanorods was quenched when the ZnS shell layer was coated onto the surfaces of the ZnO nanorods. Moreover, the ZnO-ZnS nanorods with a 17 nm-thick ZnS shell layer exhibited a substantial drop in NBE intensity when compared with other ZnO-ZnS nanorods. Coating the surfaces of the ZnO nanorods with ZnS shell layer leads to fill up the oxygen vacancies on the surface of the oxide nanorods with the S atoms. The intensity ratio of visible emission band peak to NBE peak (I_D/I_{NBE}) for the ZnO nanorods was 8.4%. The decoration of ZnS shell layer (7 nm) onto the surfaces of the ZnO nanorods decreased the NBE peak intensity and the I_D/I_{NBE} ratio was also decreased to 3.8%. Further increasing the ZnS shell layer thickness, more S atoms fill up more oxygen vacancies with an increased ZnS shell layer thickness and further decreases the I_D/I_{NBE} ratio to 2.1% and 1.8% for the ZnO-ZnS (17 nm) and ZnO-ZnS (32 nm), respectively. However, further thickening the ZnS shell layer to 46 nm, the position of deep level emission band of the ZnO-ZnS nanorods was markedly red-shifted and exhibited a substantial increase in intensity (I_D/I_{NBE} = 14.8%). A lot of surface oxygen vacancies of the ZnO nanorods have been filled up with an increased ZnS shell layer and this engendered a decreased deep level emission of the ZnO nanorods. The formation of a thicker ZnS layer (46 nm), in contrast, might form more sulfur vacancies in the ZnS crystallites with the prolonged sputtering duration, which resulted in the presence of visible deep level emission that originated from the ZnS shell crystallites [29]. Comparatively, an optimal ZnS shell layer thickness of approximately 17 nm was found to make ZnO-ZnS nanorods showing superior photoexcited charge separation efficiency between the ZnO-ZnS heterointerface.

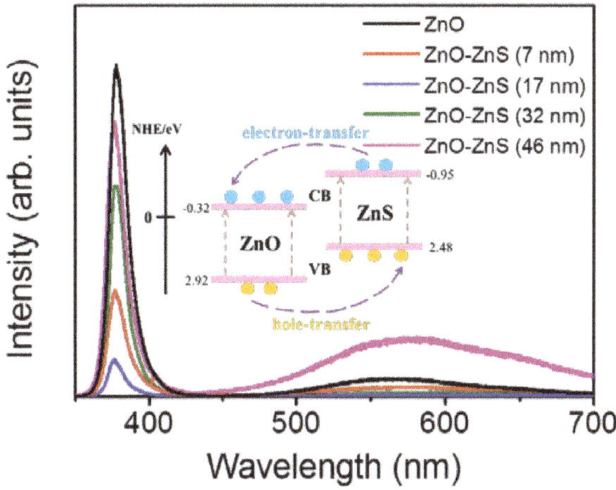

Figure 4. PL spectra of ZnO-ZnS composite nanorods with various sputtering ZnS thicknesses.

Figure 5a–e display photoresponse behaviors of the ZnO nanorods and various ZnO-ZnS core-shell nanorods. The measured photocurrents increased as a function of time after the sample was exposed to solar light, and successfully reached the steady-state value. The measured photocurrents of the sample decreased over time in the absence of solar light. The ratio of solar light irradiated currents (I_{on}) to dark currents (I_{off}) is used to determine the photoresponse performance of the composite nanorods [30]. Notably, the photoresponse of the ZnO nanorods is approximately 4.4. A substantial enhancement of the photoresponse of the ZnO nanorods was observed for the surface decoration with a ZnS shell layer. The conduction band energy level of ZnS is found to be energetically slightly higher than that of ZnO [31,32]. A possible band alignment between the ZnO and ZnS was shown in the inset of Figure 4 [33]. When the ZnO-ZnS nanorods were exposed to solar light, a large number of photoexcited electron-hole pairs are formed in both ZnS and ZnO. The electrons are readily from the conduction band of ZnS to the conduction band of ZnO. The well crystalline ZnO nanorods provide a path for transporting electrons to the adjacent electrodes and an enhanced photoresponse was observed for the ZnO nanorods coated with the ZnS shell layer. Similar photoresponse enhancement mechanism has been reported in ZnO-Cu$_2$O and ZnS-SnO$_2$ heterostructure systems, wherein a similar band alignment structure exists between the counterparts [34,35]. The photoresponse of the ZnO-ZnS nanorods with the 7 nm-thick ZnS shell layer is approximately 138. An increase of the ZnS shell layer thickness to 17 nm makes the ZnO-ZnS nanorods exhibit the highest photoresponse of approximately 358. However, further thickening the ZnS shell layer to 32 nm and 46 nm, by contrast, decreased the photoresponses of the ZnO-ZnS composite nanorods to 34 and 18, respectively. The photoresponse of nanorods is largely affected by the number of surface adsorbed O$_2$ and H$_2$O molecules [36]. The shell layer thickness has been shown to affect photocurrent size of the heterostructure systems. It has been shown in the ZnO-TiO$_2$ composite nanorods with various TiO$_2$ shell layer thicknesses, an increased coating cycle number of TiO$_2$ shell layer engenders the photoexcited electrons produced in the shell are trapped by the adsorbed oxygen molecules on the surface of TiO$_2$, and thus could not be transferred to ZnO. This results in a large drop in the photocurrent in the ZnO-TiO$_2$ system [37]. In addition, the ZnO-ZnS core-shell nanorods herein also display shorter response time and recovery time than those of pure ZnO nanorods. The response time was 68 s and the recovery time was 108 s for the ZnO nanorods. However, the response time and recovery time of the ZnO-ZnS (17 nm) was markedly shortened to 30 s and 78 s, respectively. This is associated with the charges combined with type-II band structure between ZnO and ZnS, which can easily separate the photoexcited electrons and holes into the ZnO core and ZnS shell, respectively. Notably, the PL analysis result revealed that the charge separation efficiency of the ZnO-ZnS was substantially improved with the 17 nm-thick ZnS shell layer; this is correlated with the highest photoresponse of the ZnO-ZnS (17 nm) herein.

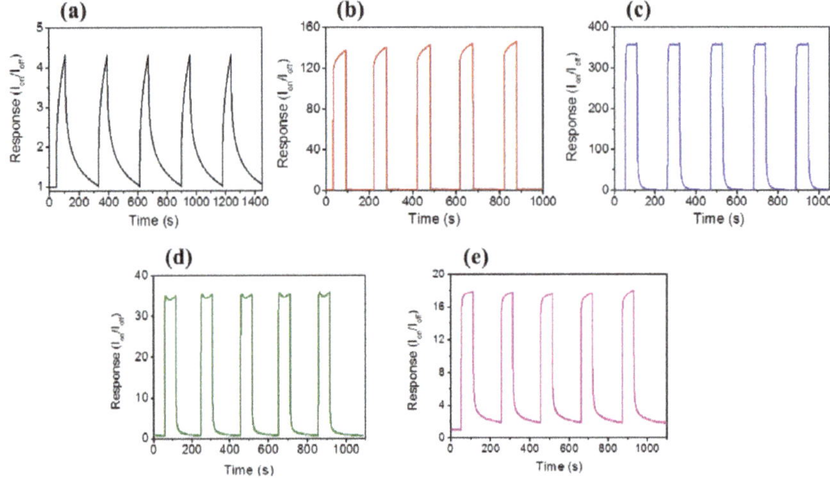

Figure 5. Photoresponse curves of various nanorods: (**a**) Pure ZnO; (**b**)ZnO-ZnS (7 nm); (**c**) ZnO-ZnS (17 nm); (**d**)ZnO-ZnS (32 nm); (**e**) ZnO-ZnS (46 nm).

The photocatalytic performance of ZnO-ZnS nanorods with various ZnS shell layers under light irradiation was further investigated. The time-course absorbance spectra for an aqueous MB solution after photodegradation containing various ZnO-ZnS nanorod samples are displayed in Figure 6a–d. The visible and intense peaks of the absorbance spectra at approximately 663 nm were due to the monomeric MB; the intensity of absorbance spectra decreased with irradiation time. This result indicates that MB dyes are gradually degraded under irradiation. The ratio of the remaining MB concentration (C) after light irradiation to the initial MB concentration without light irradiation (C$_o$) i.e., C/C$_o$ was further used to determine the photodegradation level of the MB solution containing various nanorod samples. The C/C$_o$ value was evaluated from the absorbance spectra intensity ratio at 663 nm before and after the MB solution was subjected to irradiation with various nanorod samples. The C/C$_o$ versus irradiation duration results for the MB solution containing various nanorod samples are shown in Figure 6e. Notably, dark adsorption tests were performed for various durations before the photocatalytic degradation tests under irradiation. The MB concentration changes in dark conditions are influenced by the adsorption of MB dyes on the surfaces of various nanorod samples. Notably, photoexcited electrons or holes in the ZnO-ZnS nanorods were transferred to the active surface and join in the redox reactions. The producing hydroxyl radicals were strong oxidizing agents and effectively decomposed the MB dyes [38]. Under the given irradiation duration, the complex

ZnO-ZnS (17 nm) nanorods show much greater photocatalytic activity in the decomposition of MB when compared to other ZnO-ZnS nanorods. Almost 90% of the MB solution containing ZnO-ZnS (17 nm) nanorods was discolored under irradiation for 60 min. However, only 75%, 60%, 52%, and 40% MB solution were discolored containing ZnO-ZnS (7 nm), ZnO-ZnS (32 nm), ZnO-ZnS (46 nm), and ZnO nanorods, respectively, under irradiation for 60 min. Further extended the irradiation duration to 75 min, the discoloration level of MB solution containing ZnO-ZnS (17 nm) reached a balance value of 90%, revealing that most MB dyes in the solution were degraded in the presence of the ZnO-ZnS (17 nm) nanorods under irradiation after 60 min. By contrast, the discoloration level of the MB solution containing other nanorods was further increased and did not reach a balance value. The complex ZnO-ZnS (17 nm) nanorods exhibited the suitable shell thickness of ZnS to enhance electron–hole separations at the interface of ZnO-ZnS, resulting in the highest photocatalytic efficiency. The similar improvement in photocatalytic efficiency and stability of the SnO_2 has been achieved in coupling SnO_2 with ZnS to form a heterostructure. The efficient separation of hole–electron pairs at the SnO_2-ZnS heterointerface has posited to suppress photoexcited charge recombination and inhibit photocorrosion of SnO_2-ZnS nanocomposites [39]. The PL results and photoresponse performance of the ZnO-ZnS (17 nm) nanorods supported the superior photocatalytic activity of the ZnO-ZnS (17 nm) nanorods, herein. Notably, the substantial decrease in the absorbance peak intensity in the UV region is associated with the mineralization of the MB dyes during the photodegradation process [7]. The trend that the larger intensity decrease level of the absorbance peak in the UV region for the MB solution containing ZnO-ZnS (17 nm) nanorods at the given irradiation duration is in accordance with the observations on discoloration degree of the MB solution containing various nanorod samples. The stability and reusability of the ZnO-ZnS (17 nm) nanorods were evaluated by performing recycling reactions three times for the photodegradation of the MB solution under irradiation. In Figure 6f, after three cycles, the ZnO-ZnS (17 nm) nanorods maintain the high reusability and stability; an approximately 90% photodegragation of the MB solution was observed. The ZnO-ZnS (17 nm) nanorods are advantageous in the applications of photocatalyst for organically polluted water treatment.

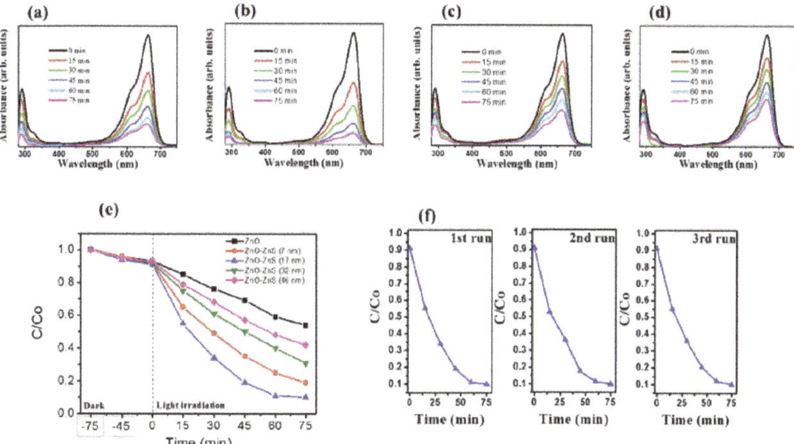

Figure 6. Intensity variation of absorbance spectra of MB solution vs. degradation duration containing various ZnO-ZnS nanorods with different sputtering ZnS thicknesses: (**a**) 7 nm; (**b**) 17 nm; (**c**) 32 nm; (**d**) 46 nm; (**e**) C/C_o vs. irradiation time for MB solution containing various ZnO-ZnS nanorods in dark conditions and under solarlight illumination. For comparison those of pure ZnO nanorods are also shown in the plot. (**f**) Recycled photodegradation performances (three test runs) of MB solution containing the ZnO-ZnS (17 nm) nanorods.

4. Conclusions

We have effectively demonstrated the photocatalytic properties of shell thickness controlled ZnO-ZnS core-shell nanorods, formed through sputtering ZnS thin films with different sputtering durations onto the surfaces of ZnO nanorod templates. The crystal structure analyses revealed that the shell layer is well crystallized, with preferring growth direction of ZnS (002). This indicates that hexagonal ZnO crystal provides well crystallographic accommodation for growth of hexagonal ZnS shell layers by sputtering deposition at 460 °C, herein. Among various ZnS shell layer thicknesses, the ZnO-ZnS nanorods with the 17 nm-thick ZnS shell layer exhibited the highest photoexcited charge separation efficiency; therefore, displayed the highest photodegradation level to MB dyes. Moreover, the ZnO-ZnS (17 nm) nanorods maintained the high reusability and stability after three cycles degradation. The results herein imply that the control of sputtering deposited ZnS shell layer thickness and crystallinity is an efficient approach to optimize the charge separation efficiency of ZnO-ZnS core-shell nanorods, and, therefore, improving photocatalytic activity of the composite nanorods.

Acknowledgments: This work is supported by the Ministry of Science and Technology of Taiwan (Grant No. MOST 105-2628-E-019-001-MY3).

Author Contributions: Yuan-Chang Liang drafted the manuscript and designed the experiments. Ya-Ru Lo, Chein-Chung Wang and Nian-Cih Xu carried out the sample preparations, material characterization analyses and data collection. All authors read and approved the final manuscript.

References

1. Wang, Z.L. Nanostructures of zinc oxide. *Mater. Today* **2004**, *7*, 26–33. [CrossRef]
2. Qi, K.; Cheng, B.; Yu, J.; Ho, W. Review on the improvement of the photocatalytic and antibacterial activities of ZnO. *J. Alloy Compd.* **2017**, *727*, 792–820. [CrossRef]
3. Singh, S.; Barick, K.C.; Bahadur, D. Shape-controlled hierarchical ZnO architectures: Photocatalytic and antibacterial activities. *CrystEngComm* **2013**, *15*, 4631–4639. [CrossRef]
4. Chang, J.; Waclawik, E.R. Facet-controlled self-assembly of ZnO nanocrystals by non-hydrolytic aminolysis and their photodegradation activities. *CrystEngComm* **2012**, *14*, 4041–4048. [CrossRef]
5. Al-Sabahi, J.; Bora, T.; Al-Abri, M.; Dutta, J. Controlled defects of zinc oxide nanorods for efficient visible light photocatalytic degradation of phenol. *Materials* **2016**, *9*, 238. [CrossRef] [PubMed]
6. İkizler, B.; Peker, S.M. Synthesis of TiO_2 coated ZnOnanorod arrays and their stability in photocatalytic flow reactors. *Thin Solid Films* **2016**, *605*, 232–242. [CrossRef]
7. Liang, Y.C.; Chung, C.C.; Lo, Y.J.; Wang, C.C. Microstructure-dependent visible-light driven photoactivity of sputtering-assisted synthesis of sulfide-based visible-light sensitizer onto ZnONanorods. *Materials* **2016**, *9*, 1014. [CrossRef] [PubMed]
8. Liang, Y.C.; Lung, T.W.; Xu, N.C. Photoexcited Properties of Tin Sulfide Nanosheet-Decorated ZnONanorodHeterostructures. *Nanoscale Res. Lett.* **2017**, *12*, 258. [CrossRef] [PubMed]
9. Rouhi, J.; Ooi, C.H.R.; Mahmud, S.; Mahmood, M.R. Facile synthesis of vertically aligned cone-shaped ZnO/ZnS core/shell arrays using the two-step aqueous solution approach. *Mater. Lett.* **2015**, *147*, 34–37. [CrossRef]
10. Ummartyotin, S.; Infahsaeng, Y. A comprehensive review on ZnS: From synthesis to an approach on solar cell. *Renew. Sustain. Energy Rev.* **2016**, *55*, 17–24. [CrossRef]
11. Lee, G.J.; Wu, J.J. Recent developments in ZnSphotocatalysts from synthesis to photocatalytic applications—A review. *Powder Technol.* **2017**, *318*, 8–22. [CrossRef]
12. Yu, L.; Chen, W.; Li, D.; Wang, J.; Shao, Y.; He, M.; Wang, P.; Zheng, X. Inhibition of photocorrosion and photoactivity enhancement for ZnO via specific hollow ZnO core/ZnS shell structure. *Appl. Catal. B Environ.* **2015**, *164*, 453–461. [CrossRef]
13. Li, W.; Song, G.; Xie, F.; Chen, M.; Zhao, Y. Preparation of spherical ZnO/ZnS core/shell particles and the photocatalytic activity for methyl orange. *Mater. Lett.* **2013**, *96*, 221–223. [CrossRef]

14. Zhang, K.; Jing, D.; Chen, Q.; Guo, L. Influence of Sr-doping on the photocatalytic activities of CdS–ZnS solid solution photocatalysts. *Int. J. Hydrog. Energy* **2010**, *35*, 2048–2057. [CrossRef]

15. Zhang, R.; Wang, B.; Wei, L.; Li, X.; Xu, Q.; Peng, S.; Kurash, I.; Qian, H. Growth and properties of ZnS thin films by sulfidation of sputter deposited Zn. *Vacuum* **2012**, *86*, 1210–1214. [CrossRef]

16. Wu, X.; Lai, F.; Lin, L.; Lv, J.; Zhuang, B.; Yan, Q.; Huang, Z. Optical inhomogeneity of ZnS films deposited by thermal evaporation. *Appl. Surf. Sci.* **2008**, *254*, 6455. [CrossRef]

17. Krasnov, A.N.; Hofstra, P.G.; McCullough, M.T. Parameters of vacuum deposition of ZnS:Mn active layer for electroluminescent displays. *J. Vac. Sci. Technol. A* **2000**, *18*, 671. [CrossRef]

18. Patel, S.P.; Chawla, A.K.; Chandra, R.; Prakash, J.; Kulriya, P.K.; Pivin, J.C.; Kanjilal, D.; Kumar, L. Structural phase transformation in ZnSnanocrystalline thin films by swift heavy ion irradiation. *Solid State Commun.* **2010**, *150*, 1158–1161. [CrossRef]

19. Roy, P.; Ota, J.R.; Srivastava, S.K. Crystalline ZnS thin films by chemical bath deposition method and its characterization. *Thin Solid Films* **2006**, *515*, 1912–1917. [CrossRef]

20. Liang, Y.C.; Deng, X.S. Structure dependent luminescence evolution of c-axis-oriented ZnOnanofilms embedded with silver nanoparticles and clusters prepared by sputtering. *J. Alloy Compd.* **2013**, *569*, 144–149. [CrossRef]

21. Liang, Y.C.; Deng, X.S.; Zhong, H. Structural and optoelectronic properties of transparent conductive c-axis-oriented ZnO based multilayer thin films with Ru interlayer. *Ceram. Int.* **2012**, *38*, 2261–2267. [CrossRef]

22. Bösenberg, U.; Buchmann, M.; Rettenmayr, M. Initial transients during solid/liquid phase transformations in a temperature gradient. *J. Cryst. Growth* **2007**, *304*, 275–280. [CrossRef]

23. Liang, Y.C.; Lee, H.Y.; Liang, Y.C.; Liu, H.J.; Wu, K.F.; Wu, T.B. Surface evolution and dynamic scaling of heteroepitaxial growth of (La,Ba)MnO$_3$ films on SrTiO$_3$ substrates by rf magnetron sputtering. *Thin Solid Films* **2006**, *494*, 196–200. [CrossRef]

24. Liang, Y.C.; Liang, Y.C. Effects of lattice modulation on magnetic properties of epitaxial ferromagnetic/paramagnetic manganite superlattices. *J. Cryst. Growth* **2006**, *296*, 104–109. [CrossRef]

25. Pathak, T.K.; Kumar, V.; Purohit, L.P.; Swart, H.C.; Kroon, R.E. Substrate dependent structural, optical and electrical properties of ZnS thin films grown by RF sputtering. *Physica E* **2016**, *84*, 530–536. [CrossRef]

26. Liang, Y.C.; Lee, H.Y. Growth of epitaxial zirconium-doped indium oxide (222) at low temperature by rf sputtering. *Crystengcomm* **2010**, *12*, 3172–3176. [CrossRef]

27. Cao, G.; Yang, H.; Hong, K.; Hu, W.; Xu, M. Synthesis of long ZnO/ZnS core–shell nanowires and their optical properties. *Mater. Lett.* **2015**, *161*, 278–281. [CrossRef]

28. Chen, P.; Gu, L.; Cao, X. From single ZnO multipods to heterostructured ZnO/ZnS, ZnO/ZnSe, ZnO/Bi$_2$S$_3$ and ZnO/Cu$_2$S multipods: Controlled synthesis and tunable optical and photoelectrochemical properties. *CrystEngComm* **2010**, *12*, 3950–3958. [CrossRef]

29. Sadollahkhani, A.; Nur, O.; Willander, M.; Kazeminezhad, I.; Khranovskyy, V.; Eriksson, M.O.; Yakimova, R.; Holtz, P.O. A detailed optical investigation of ZnO@ZnS core–shell nanoparticles and their photocatalytic activity at different pH values. *Ceram. Int.* **2015**, *41*, 7174–7184. [CrossRef]

30. Liang, Y.C.; Liao, W.K. Annealing induced solid-state structure dependent performance of ultraviolet photodetectors made from binary oxide-based nanocomposites. *RSC Adv.* **2014**, *4*, 19482–19487. [CrossRef]

31. Daghrir, R.; Drogui, P.; Robert, D. Modified TiO$_2$for environmental photocatalytic applications: A review. *Ind. Eng. Chem. Res.* **2013**, *52*, 3581–3599. [CrossRef]

32. Sang, H.X.; Wang, X.T.; Fan, C.C.; Wang, F. Enhanced photocatalytic H$_2$ production from glycerol solution over ZnO/ZnS core/shell nanorods prepared by a low temperature route. *Int. J. Hydrog. Energy* **2012**, *37*, 1348–1355. [CrossRef]

33. Xitao, W.; Rong, L.; Kang, W. Synthesis of ZnO@ZnS–Bi$_2$S$_3$ core–shell nanorod grown on reduced graphene oxide sheets and its enhanced photocatalytic performance. *J. Mater. Chem. A* **2014**, *2*, 8304–8313. [CrossRef]

34. Liu, X.; Du, H.; Wang, P.; Lim, T.T.; Sun, X.W. A high-performance UV/visible photodetector of Cu$_2$O/ZnO hybrid nanofilms on SWNT-based flexible conducting substrates. *J. Mater. Chem. C* **2014**, *2*, 9536–9542. [CrossRef]

35. Huang, X.; Yu, Y.Q.; Xia, J.; Fan, H.; Wang, L.; Willinger, M.G.; Yang, X.P.; Jiang, Y.; Zhang, T.R.; Meng, X.M. Ultraviolet photodetectors with high photosensitivity based on type-II ZnS/SnO$_2$ core/shell heterostructured ribbons. *Nanoscale* **2015**, *7*, 5311–5319. [CrossRef] [PubMed]

36. Li, Q.H.; Gao, T.; Wang, Y.G.; Wang, T.H. Adsorption and desorption of oxygen probed from ZnO nanowire films by photocurrent measurements. *Appl. Phys. Lett.* **2005**, *86*, 509. [CrossRef]
37. Panigrahi, S.; Basak, D. Core–shell TiO$_2$@ZnO nanorods for efficient ultraviolet photodetection. *Nanoscale* **2011**, *3*, 2336–2341. [CrossRef] [PubMed]
38. Liang, Y.C.; Lung, T.W. Growth of hydrothermally derived CdS-Based nanostructures with various crystal features and photoactivated properties. *Nanoscale Res. Lett.* **2016**, *11*, 264. [CrossRef] [PubMed]
39. Hu, L.; Chen, F.; Hu, P.; Zou, L.; Hu, X. Hydrothermal synthesis of SnO$_2$/ZnSnanocomposite as a photocatalyst for degradation of Rhodamine B under simulated and natural sunlight. *J. Mol. Catal. A Chem.* **2016**, *411*, 203–213. [CrossRef]

MDPI
St. Alban-Anlage 66
4052 Basel
Switzerland
Tel. +41 61 683 77 34
Fax +41 61 302 89 18
www.mdpi.com

Materials Editorial Office
E-mail: materials@mdpi.com
www.mdpi.com/journal/materials

CPSIA information can be obtained
at www.ICGtesting.com
Printed in the USA
LVHW072356160919
631222LV00003B/45/P